CLIMATE CHANGE

A Convenient Truth

D0912221

Jim Hollingsworth

ISBN 978-1-64559-746-9 (Paperback)
ISBN 978-1-64559-745-2 (Digital)

Covenant Books, Inc.
11661 Hwy 707
Murrells Inlet, SC 29576
www.covenantbooks.com

To my grandchildren, who will have to live
with the policy decisions we make.

THANK YOU

I wish to thank the following professionals in climate change or related who read the book and made many helpful changes:

Tom DeWeese
Joe Bastardi
Bob Zybach
Caleb Rossiter
Patrick Michaels
Joe Tomlinson
Allen Rogers
Ernst Motte
Richard Meyer
Tammy Meatzie
Russ Fulcher
Tom Sheahan
Katie Coble

ENDORSEMENTS

Jim Hollingworth's book, *Climate Change: A Convenient Truth*, is a breezily-written short book that is worthwhile to anyone who is curious about this "global warming/climate change" phenomenon that has been sweeping Western Civilization since the 1980s and has now become a multi-trillion-dollar industry. The book comprises forty-eight (very short) chapters, plus seven Appendices (for anyone who cares to read more in-depth on the topic). Its nonintimidating, simplistic style should not offend anyone—whether you're in middle school or college or have been away from school for fifty years.

—Daniel W. Nebert, BA [biochem], MS [bio-
phys], MD [pediatrics/genetics]
Department of Pediatrics & Molecular Developmental
Biology, Division of Human Genetics,
Cincinnati Children's Hospital, Cincinnati, OH 45229
Consultant, Department of Anesthesia,
Stanford University Medical Center
Stanford, CA 94305

As a litigation attorney for nearly thirty years, I know a lot about evidence. The evidence presented for "climate change" simply does not add up. There is no basis for saying the science is settled or 97 percent of scientists support global warming theories. Many highly qualified scientists have shown that there is no threat from carbon dioxide. Jim Hollingsworth shows the valid evidence. His book is an important resource for those who want to know what the science really shows.

—Richard Botteri, Attorney at Law

"Hidden in this small book is all the information one needs to fully understand the frauds, which are being perpetuated on society by environmental extremists desiring to foist their miserable personalities upon an unsuspecting public. Hollingsworth moves from the now exposed acid rain fraud of yesteryear to the biggest fraud ever perpetrated on society; man caused Climate Change. This deception has now cost the nations of our planet, well over a trillion dollars, in the belief that man is more powerful than Nature, and thus can set the Earth's thermostat.

Without ever getting into the weeds of math and science, anyone with a tenth-grade education will be able to understand precisely how they have been hoodwinked these many years. A four-hour investment in reading time will yield multiple eureka moments.

—JAY LEHR, PhD
Senior Policy Analyst for the International
Climate Science Coalition
and Former Science Director of The Heartland Institute

Climate Change: A Convenient Truth is a vital, complete, and above all, truthful presentation on climate science. Science must pursue the untainted truth. This book dispels the confusion that exists about "climate change" because politics and money, rather than science, are involved in promoting anthropogenic global warming. Regardless of your background, you will benefit from learning the truth about climate science.

—Bernard "Bernie" Kepshire, PhD, MS, BA
Fisheries biologist; background includes Alaska
Department of Fish and Game, Oregon Department
of Fish and Wildlife, and Oregon State University

This book is much needed to give people an impartial view of what is really happening to our climate and what is not happening. All of us taxpayers are, in fact, funding billions and soon trillions of dollars to be spent on "decarbonizing" our atmosphere, even though CO_2 is food for all plants. I hope that many people start asking more questions about the logic of today's energy policies after having read Jim's well-written book.

—Lars Schernikau, PhD, Energy Economics

My good friend, Jim Hollingsworth, has written a very accessible book on the sham, scam, and hoax of global warming hysteria. Folks, there is nothing to be afraid of, and Jim's informative book will settle whatever fears you may have. Give it a read—you won't be disappointed!

—Bryan Fischer, Host of "Focal Point" on the American Family Radio Network

Jim Hollingsworth's new book, *Climate Change: A Convenient Truth*, lays it all out—in short, readable chapters that address virtually every fear-inducing claim concerning climate change—and finds them wanting. If you're worried about dangerous man-made climate change, but still have an open mind, this is a must read. If you're already skeptical, but want an easy-to-use reference guide, this is what you've been looking for.

—Paul Driessen,
Author of Eco-Imperialism: Green Power-Black Death
and many articles and books on energy, climate change,
human rights and the dark side of renewable energy

Finally, a book on climate that anyone can understand. If you master the forty-eight short chapters in this little book you will be well on your way to understanding the basics of climate change: Man's part and Nature's. This is a book that needs to be read more than once and then shared with your friends. Careful reading should convince you that even a majority can be wrong.

CONTENTS

INTRODUCTION

A lot has been written about global warming/climate change, especially in the last couple of years. We have been led to believe that if we do not take immediate, expensive and decisive action the world as we know it will come to a dramatic end.

But, is it true? Where is the evidence that man is the cause of this warming? Can we do anything about it, or do we just let nature have its way?

No-one can dispute the fact that the earth is warming. It has continued to warm since the end of the last ice age, about 10,000 years ago. At one time North Idaho was buried under a mile of ice.

But what is the evidence that man is the chief cause of this warming, or that we can stop a further increase in temperature simply by ending our use of all fossil fuels (coal, oil, natural gas, etc.)?

This book was written for you. It is time that you had a chance to know the truth. The mainstream media wants to convince you that we are headed for a catastrophe if we simply do nothing. Twelve years ago we were told we only had ten years left, yet those ten years have come and gone without any change in our climate.

There are three types of people who consider climate change.

The first is the government paid scientists and those associated with them who will not be convinced no matter what is said, because if they had to admit they were wrong it would end their employment.

The second are those who only watch television, never pick up a book and really do not want to know the truth until their house goes dark and it is too late to do anything about it.

The third group includes a large number of people who have received much conflicting information and still do not know the truth, yet they seek to understand it.

So, in which group are you? This book was written for the third group. It is written so you can understand climate change and how it affects you. You have to understand that the program of the first group has to do with a lot more than just climate. You can have no effect on them unless you know the truth.

You can seek to know the truth, or you can continue to ignore it. It is important, however, that each one of us not only knows the truth, but does something about it to get our government to take a stand on doing what is right. This book details most of the important issues of climate change.

Finally, are you going to believe everything put out concerning the Green New Deal and similar programs, or will you take a couple of hours to read this book and then do something about it? I trust you will.

CHAPTER 1

97 Percent Agreement

President Obama and Secretary of State John Kerry have both made the statement that 97 percent of scientists agree about global warming; it is man-caused and it is dangerous. Unfortunately, science does not depend on agreement but on proven theories. Scientists work to disprove theories, not to prove them. It is unfortunate that many scientists come with a preconceived notion about a particular aspect of science, rather than seeking the facts and seeing where they lead.

That is the problem with this 97 percent. When you actually look at the data, you find it is far less than 97 percent.

If you look at the literature, the specific meaning of the 97 percent claim is 97 percent of climate scientists agree that there is a global warming trend and that human beings are the main cause, that is, that we are over 50 percent responsible. The warming is a whopping 0.8 degrees over the past 150 years, a warming that has tapered off to essentially nothing in the last decade and a half.

Here is what John Kerry said:

> And let there be no doubt in anybody's mind that the science is absolutely certain…97 percent of climate scientists have confirmed that climate change is happening and that human activity is responsible…they agree that, if we continue to go down the same path that we are going down

today, the world as we know it will change—and
it will change dramatically for the worse. (Alex
Epstein)

One of the authors of the concept of 97 percent is John Cook
who did a survey on the subject. The problem is that not all the scientists who were surveyed agreed with the Cook premise. Look him
up and other papers about the 97 percent consensus.

Here is Cook's summary of his paper: "Cook et al. (2013) found
that over 97 percent [of papers he surveyed] endorsed the view that
the Earth is warming up and human emissions of greenhouse gases
are the main cause."

But even a quick scan of the paper reveals that this is not the
case. Cook is able to demonstrate only that a relative handful endorse
"the view that the Earth is warming up and human emissions of
greenhouse gases are the main cause."

When the study was publicly challenged by economist David
Friedman, one observer calculated that only 1.6 percent explicitly
stated that man-made greenhouse gases caused at least 50 percent of
global warming.

The 97 percent claim is a deliberate misrepresentation designed
to intimidate the public—and numerous scientists whose papers
were classified by Cook. [1]

Many of the scientists whose papers Cook quoted said that their
data was grossly misrepresented. So, the 97 percent is closer to 1.6
percent.

Most of the people who quote the 97 percent have no idea what
that figure represents. Here is just one example, Senator Ted Cruz
interviewing Aaron Mair of the Sierra Club.[2]

Now here is something else that is at least worth considering—
the *Petition Project* in which 31,487 scientists signed the petition and

---

[1]   Alex Epstein
      http://www.forbes.com/sites/alexepstein/2015/01/06/97-of-climate-scientists-
      agree-is-100-wrong/#4ea02dc13f9f
[2]   https://www.youtube.com/watch?v=MSmZ9c_15j0

9,024 of them with PhDs in their field. Here is what they were agreeing to when they signed the petition.

Petition

We urge the United States government to reject the global warming agreement that was written in Kyoto, Japan in December, 1997, and any other similar proposals. The proposed limits on greenhouse gases would harm the environment, hinder the advance of science and technology, and damage the health and welfare of mankind.

There is no convincing scientific evidence that human release of carbon dioxide, methane, or other greenhouse gases is causing or will, in the foreseeable future, cause catastrophic heating of the Earth's atmosphere and disruption of the Earth's climate. Moreover, there is substantial scientific evidence that increases in atmospheric carbon dioxide produce many beneficial effects upon the natural plant and animal environments of the Earth.

_____   ☐ Please send more petition cards for me to distribute.
          Please Sign Here

My academic degree is B.S.☐ M.S.☐ Ph.D.☐ in the field of _____

	I have specialized scientific experience in:
_____ Name	
_____ Street	
_____ City, State, and Zip	

---

www.petitionproject.org

Go to the Web site to see a lot more information including lists of those who signed the petition. You may even see someone you know. The very first person to sign the petition was Dr. Edward Teller, sometimes called "the father of the atomic bomb."

The point is well taken: No matter their 97 percent consensus, the truth is that there are a number of climatologists and other scientists who do not agree with Obama, Kerry, and others. What is important is not the number who may agree on the subject, but what is the truth of the matter. Many of those scientists who reject the view of the alarmists are listed at the end of this book.

Man is not the main cause of global warming. As always, nature and the sun have much more control over our weather than man does.

CHAPTER 2

100 Percent Renewable Energy

The prospect of using 100 percent renewable energy for electricity, heating and cooling, and transportation is motivated by the so-called global warming pollution and other environmental issues. But to shift our entire energy system to 100 percent renewable requires a total change of how we live and do business. Although renewable energy is increasing, it is far away from being a totally renewable program.

In 2013, the Intergovernmental Panel on Climate Change (IPCC) said that powering the nation with 100 percent renewable was not only possible but practical. However, to do so would mean covering huge areas of America with solar panels and wind machines.

To begin with, we need to understand the magnitude of the challenge in changing over to 100 percent renewable. Here is a basic table of how our electricity is presently generated:

Natural gas	32.1%
Coal	29.9%
Nuclear	20%
Hydro	7.4%
Wind	6.3%
Solar	1.3%
Petroleum	.5%
Other gases	.3%
Others	2.2%

Most sources talk about wind and solar as the sources for power with 100 percent renewable. You can see it would take a lot more equipment to make that much electricity, and as mentioned above, it would mean covering much of America with wind and solar installations. According to the chart above, wind and solar make up only about 7.6 percent of our electricity, and I suspect that even this figure is high.

It is possible to produce 100 percent renewable with many more nuclear power plants, but there is presently a lot of resistance against nuclear, and there is also a movement to remove the dams to make fish migration easier. If all these things happen, we may have to go back to living as the frontier pioneers did when they first came to America, or America may simply go dark.

It is also important to keep in mind that both wind and solar have serious problems. Wind only generates when the wind blows, and solar does not generate at night or when it is cloudy. To overcome this problem, another generator must be kept idling and prepared to go online with a moment's notice, because it takes several hours to start one otherwise. Usually, these are coal or natural gas-fired generators. But when idling, they use almost as much fuel as they would under full load, and in doing so, they produce almost the same amount of carbon dioxide, so you really have not accomplished anything. Wind is also deadly for many species of threatened birds such as eagles, hawks, and bats.

The presentations of programs of 100 percent renewable are generally guilty of just not having thought the science through very thoroughly to understand all the challenges associated with their operation.

Even with 100 percent renewable energy, global warming will not be stopped if it can be shown that a major portion of temperature rise has natural causes. Further, man is not the chief cause of global warming.

CHAPTER 3

Acid Rain

Acid rain is a rain or any other form of precipitation that is unusually acidic, meaning that it has elevated levels of hydrogen ions. It can have harmful effects on plants, aquatic animals, and infrastructure. Acid rain is caused by emissions of carbon dioxide, sulfur dioxide, and nitrogen oxide, which react with the water molecules in the atmosphere to produce acids. Some countries have made efforts since the 1970s to reduce the release of sulfur dioxide and nitrogen oxide into the atmosphere with positive results.

The pH scale is between 0 and 14 with 7 being pure water and neutral. Less than 7 is acidic, and greater than 7 is basic or alkaline. However, natural, unpolluted rainwater actually has a pH of about 5.6 (acidic). Rainwater is acidic because of CO_2 (carbon dioxide), NO_2 (nitric oxide), and SO_2 (sulfur dioxide) in the air.

Carbon dioxide at over 410 parts per million (2019) makes the greatest contribution to the acidity of rainwater, though it is a very small part of our atmosphere. These acid-causing gases come from industry, lightning, and volcanoes.

All three of these gases react with water to form weak acids, and this is what makes rainwater acidic at times and in some locations.

In some areas of the United States, the pH of rainwater can be 3.0 or lower, approximately one thousand times more acidic than normal rainwater. In 1982, a fog on the West Coast of the United States measured a pH of 1.8. Rainwater that is high in acids can kill fish, damage crops, and erode buildings and monuments because they are made of limestone and marble, which dissolve in acid.

Although rainwater can be affected by these compounds which can produce acids, most rainwater is not very acidic.

Although carbon dioxide can produce acid in rain, sulfur dioxide is worse as it produces sulfuric acid (H_2SO_4).

Most of the industrial causes of sulfur dioxide have been controlled so that there are very few places that are now producing acid rain today.

In areas with volcanoes, there is an increase in acid rain, but the rain that often accompanies the eruptions tends to clean the acid from the air. Man is actually the cause of very little acid rain.[3]

---

[3] http://www.chemistry.wustl.edu/~edudev/LabTutorials/Water/FreshWater/
acidrain.html,
https://en.wikipedia.org/wiki/Acid_rain

CHAPTER 4

Albert (Al) Gore

When Al Gore screams out in Senate testimony, "The Earth has a fever," he is not so much developing a scientific theory as he is seeking to use fear to drive a political agenda (*The Myth of Global Warming*, Jim Hollingsworth; see Appendix).

Al Gore wrote a book and later a movie, *An Inconvenient Truth*. There were at least thirty-five errors of fact in them. A Mr. Stewart Dimmock of England sued to prevent the film being shown in the schools. Judge Michael Burton ruled that the film could be shown, but the teacher had to show the errors in the film. The judge listed nine (9) errors:

1. Mr. Gore claimed the sea would rise up to twenty feet, caused by the melting of part of Antarctica and Greenland. The judge said this was alarmist and this would actually take several thousand years.
2. Pacific atolls were being inundated because of global warming, but the judge ruled there was no evidence that this was causing evacuations.
3. The film talked about shutting down the "ocean conveyor" which carries heat from the tropics to the Arctic and Antarctic, but the judge ruled that this would be very unlikely to happen in the near future.

4. Mr. Gore showed two graphs which plotted temperature rise and rise in carbon dioxide which showed "an exact fit," but the judge indicated that "the two graphs do not establish what Mr. Gore asserts."

5. Mr. Gore insisted that the disappearance of snow from the top of Mt. Kilimanjaro in Africa was caused by global warming, but the judge ruled it has not been proven that this was caused by human-induced climate change.

6. In the film, the assertion is made that the drying up of Lake Chad was caused by global warming, but the judge ruled that this was probably caused by other factors, such as overgrazing and regional climate variability. (Rivers which fed the lake were also diverted.)

7. Mr. Gore insisted that Hurricane Katrina and the subsequent damage in New Orleans were caused by global warming, but the judge ruled there was "insufficient evidence to show that."

8. Mr. Gore said that some polar bears had drowned after swimming long distances, but the judge said it was more likely because of a storm. (Polar Bears can actually swim very long distances.)

9. Mr. Gore insisted that the bleaching of corals was caused by global warming, but the judge felt other factors were also involved, and it is difficult to separate causes.

The judge did not ban the film from British schools but ruled that it can only be shown with guidance notes (pointing out the above errors) to prevent political indoctrination. Mr. Gore, in the film and book, insisted that humanity is sitting on a ticking time bomb. The judge did not completely condemn the film but said the idea of a ticking time bomb reflected Mr. Gore's thesis on global warming and not the real world. Judge Michael Burton indicated that errors had entered because of a context of alarmism and exaggeration.

The film (and book) insisted that if the majority of the scientists of the world were right, there are only ten years left to avert a major catastrophe that could send our planet into a tailspin of epic destruction involving extreme weather, floods, droughts, epidemics, and killer heat waves beyond anything we have ever experienced. But the judge ruled that the apocalyptic vision presented in the film was politically partisan and thus not impartial scientific analysis of climate change.[4]

It is important to point out that the ten years mentioned in the film and the book have already passed with no calamity.

I also think that whenever *An Inconvenient Truth* is shown, a companion film *The Great Global Warming Swindle* should be shown along with it. The film can be found in a number of locations online. Here is one of them:[5]

Let us now look at the other twenty-six errors that the judge did not mention.

1. 100 ppmv of CO_2 melting mile-thick ice. *At one time there was over a mile of ice in parts of the now civilized world that has melted. But for one degree change in temperature, this would take many years.*

2. Hurricane Catarina in Brazil was man-made. *This was an unusual occurrence, but when this happened, it was cooler there, not warmer.*

3. Japanese typhoons set a new record. *Actually, the number of typhoons and tropical storms has fallen over the last many years.*

4. Hurricanes are getting stronger. *Hurricane Katrina was a category five in Florida and New Orleans in 2005. Most of the damage in New Orleans was because of poor construction of levies (and remember that much of New Orleans is below sea level), but after 2005 (with Dennis, Emily, Katrina, Rita, and Wilma), there were no serious hurricanes for many years. We were told to expect more severe hurricanes in 2006, but there were no serious hurricanes again until 2017.*

---

4   http://www.telegraph.co.uk/news/earth/earthnews/3310137/Al-Gores-nine-Inconvenient-Untruths.html

5   https://www.youtube.com/watch?v=oYhCQv5tNsQ

5. Mr. Gore says big storm insurance losses are increasing, *but they are not.*

6. Mr. Gore says that flooding in Mumbai is increasing because of global warming, *but it is not.*

7. Severe tornadoes are becoming more frequent, *but they have actually decreased for at least fifty years.*

8. The sun heats the Arctic Ocean, *but it is the atmosphere through complicated energy transfers and wind that is heating the Arctic Ocean.*

9. The Arctic is warming fastest, *but the Arctic warms and cools through time. At times there is lots of ice and at other times there is lots of open water with the Northwest Passage. But the Arctic is not warming faster than the rest of the Earth.*

10. Mr. Gore insists that the Greenland ice sheet is unstable. *It is not. Greenland is not just a flat area covered with ice but has mountains and valleys. There is always some melt in summer, but Greenland actually receives a lot of snow. For the Greenland ice sheet to melt, the temperature would have to be several degrees warmer for several thousand years.*

11. Himalayan glacial melt waters are failing. *Glacial melt is not failing, and the rivers get their water mainly from new snow.* Glacial melt is the water from melting glaciers. These rivers get their water from glacial melt and also from rain and snow, just like the rest of the world.

12. Gore insists that Peruvian glaciers are disappearing. *They are not. Much of Peru has been ice-free for the past ten thousand years, except for the tallest mountains.*

13. Mountain glaciers worldwide are disappearing. *It is easy to pick and choose glaciers. Some are melting, and some are growing. As Earth has warmed since the end of the little ice age (about 1850), the glaciers have continued to melt, and this has nothing to do with the activities of humans.*

14. The Sahara Desert is drying. *One wonders how the Sahara could actually get any dryer, but recent records indicate that some of the Sahara is actually getting greener.*

15. The West Antarctic ice sheet is unstable. *Actually it is not. In much of the Antarctic, the ice is getting thicker. A couple of years ago, a team of scientists made a trip to Antarctica to prove global warming and got stuck in the ice and had to be rescued. In some places, the temperature is actually falling.*

16. Antarctic Peninsula ice shelves are breaking up. *This does not seem likely. The Antarctic Peninsula is only about 2 percent of the continent. For the rest of Antarctica, the ice is growing thicker.*

17. Larsen B Ice Shelf broke up because of global warming. *As this ice shelf has come and gone, this is, at best, very misleading.*

18. Mosquitoes are climbing to higher altitudes. *They are not. Anyone who has been to Alaska knows that mosquitoes love cold climates.*

19. Many tropical diseases spread through global warming. *This includes dengue fever, Lyme disease, West Nile virus, arenavirus, avian flu, Ebola virus, E. coli, hantavirus, legionnaires' disease, multidrug-resistant TB, Nipah virus, SARS, and Vibrio Cholerae. But it is not happening. These diseases are not spread by warmer weather but by rats and other animals.*

20. Al Gore says that West Nile virus in the United States is spread through global warming. *But the climate in the United States ranges from very warm to very cold, and this disease spreads in any climate.*

21. Carbon dioxide is pollution. *Carbon dioxide in not pollution but makes plants grow even better. Carbon dioxide is necessary for all living things, both plants and animals.* Many commercial greenhouses deliberately raise the carbon dioxide level 3–4 times higher than normal to aid in the growth of new plants.

22. The European heat wave of 2003 killed 35,000 people. *This would be hard to prove. People may die because of lack of air-conditioning. More people die in cold weather than in warm weather. Usually, warmer weather is better than cold weather, and many people move South in the winter to take advantage of warmer weather.*

23. Pied flycatchers cannot feed their young. *These depend on caterpillars which are hatching earlier. The birds have to move north to catch them. Of course, the caterpillars might turn into butterflies, so there is a balance.*

24. Gore's bogus pictures and film footage. *Throughout the film, Gore uses one section of the film to depict a flycatcher feeding her young, but she is actually a tern with a fish. There are other instances throughout the film where there is misleading footage, such as a receding glacier that is actually advancing, etc.*

25. Operators of the Thames Barrier are closing more frequently because of the ocean. *They are not doing so. There are other reasons for changing the barrier to retain water when the tide is exceptionally low.*

26. No fact is in dispute by anyone. *This simply is not true as the above list attests.*

Al Gore makes all sorts of scientific claims, but many of them are not true. When we read things in the press, we need to take time to analyze them to understand if they are actually supported by the evidence.[6]

From the evidence available it must be obvious that man is not the chief cause of global warming.

---

[6] http://www.lomborg-errors.dk/Goreacknowledgederrors.htm

CHAPTER 5

An Inconvenient Lie (Resources)

Dr. Timothy Ball has a number of resources online, some quite short (eight minutes) and some longer (fifty-eight minutes). He is also featured in some full-length films on global warming.

> Dr. Tim Ball (fifteen minutes)[7]
> Dr. Tim Ball (fifty-eight minutes)[8]

Another resource is *The Great Global Warming Swindle*; Dr. Ball is also featured in this film along with some other great scientists. [9]

Global Warming: An Inconvenient Lie (DVD album)

This DVD album features some of the world's top scientists, researchers, and journalists who soundly reject the theory of man-caused global warming. If you are wondering why there is so much pushback against the theory of man-made global warming, here is the answer. Thirteen hours of intense programming will jolt even the most adamant believers. [10]

---

[7] https://www.youtube.com/watch?v=ksMYjzWSlI4
[8] https://www.youtube.com/watch?v=_sbo8Ods8M0
[9] https://www.youtube.com/watch?v=D3Y4WmMQtaQ
[10] http://www.realityzone.com/product/inconvenientlie/

An Inconvenient Lie: The Facts (1:50 minutes)[11]
An Inconvenient Lie, Lord Christopher Monckton (1:04 minutes)[12]

---

[11] https://www.youtube.com/watch?v=Kt7EIOE7phU
[12] https://www.youtube.com/watch?v=dDNcWxvy4ds

CHAPTER 6

Antarctic Ice Shelf Melting

Antarctica is by far the coldest continent, with the coldest temperature ever recorded being -89.2°C at the Russian Vostok Station on July 21, 1983. Temperatures in the interior are often -80°C with low precipitation, making Antarctica a frozen desert. The South Pole averages less than 10 cm of snow per year. Note, those temperatures are -89.2 and -80, respectively (*Climate Change Reconsidered II: Physical Science*, p. 639).

The only part of Antarctica that has any melting is far from the South Pole. The rest is well below freezing all year long. Antarctica simply is not melting.

It is interesting to note that a couple of years ago, a group of global warming alarmists made a trip by boat to the waters around Antarctica to prove a point. The point they actually proved is that it is quite cold there; they got stuck in the ice and had to be rescued.

Scientists have found that the winds blowing off Antarctica and strong ocean currents that circle the frozen continent have a strong effect on controlling the Antarctic sea ice. They believe that this affect has more control over the ice than changes in temperature.

The Arctic loses ice because it is one big pool of water, and this affects the ice. While the Arctic is losing ice, Antarctic is gaining ice.

Keep in mind that the North Pole is all open water. Any ice that melts there does not affect the level of the ocean. (You can see this when you put an ice cube on top of a full glass of water. As the ice melts, it does not cause the glass to overflow.)

Antarctic, however, is a continent, so the only place that ice melts is around the fringes of the continent. This ice is floating just like at the North Pole.

It is obvious that anthropogenic global warming is not affecting the ice at Antarctica.

CHAPTER 7

Arctic Ice Shelf Melting

The Arctic has held an interest for mankind for a couple of centuries. Early interest was in searching for the Northwest Passage. Many ships set out to find the Passage; most of them did not come back, and the passengers often died because of their folly. The most famous was the Resolute, which was found abandoned the next summer when the ice had melted. It was found and repaired by the United States and given back to England who kept it for several years. But when it was decommissioned, the British made a desk out of part of it and gave it to the United States president, which the president is still using, the Resolute desk.

But in recent years, most of the interest in the Arctic has centered on polar bears, the icon of the climate change movement. It is believed by some that the polar bears hunt the seals, its natural food, from off the ice, and when there is no ice, the bears go hungry and starve.

There is obviously a hole in this theory because the number of polar bears has gone from about five thousand fifty years ago to the present (2019) of about 25,000. Polar bears have survived for thousands of years, so they must have learned to hunt seals on land or developed a diet for other things. They may even have roamed south into grizzly bear territory. Of course, since polar bears are great swimmers, they may have learned to catch seals in open water.

As far as the extent of the sea ice, it depends on to whom you talk. Some see the ice as increasing, while others see it as decreasing. The Arctic has a fascinating history, and there are even people who live around the Arctic Ocean.

Climate change has not had a lasting effect on the ice in the Arctic Ocean.

CHAPTER 8

Atmospheric Carbon Dioxide Increasing

Measurements of atmospheric carbon dioxide have been made for over one hundred years. Following the carbon dioxide curve is interesting as it looks something like a sawtooth. In the springtime, the carbon dioxide level is the highest. Then plants begin to grow, and the level drops. This continues all summer. Then in the fall, when the leaves fall off, the carbon dioxide level begins to rise again.

The chief monitoring station is on Mauna Loa, a mountain on the island of Hawaii. It was located up high in an effort to avoid any possible contamination from the volcanoes located below. The Kilauea volcano is on the island of Hawaii (the Big Island) and is still active, so that could affect the levels of carbon dioxide.

The level of carbon dioxide was the same from the time of about nine thousand years ago until about 1840 when it began to rise. Here are some figures:

1840	280 ppmv (parts per million by volume)
1910	300 ppmv
1958	313 ppmv
1990	350 ppmv
2018	406 ppmv
2019	410 ppmv

The difficulty when trying to tie carbon dioxide to temperature is that there were times when the carbon dioxide was increasing, but the temperature was not rising, or it was even falling. Then, careful research has shown that temperature rises first followed by a rise in carbon dioxide. So it is possible that one is not driving the other, but they are being driven by some other source.

Careful analysis also indicated that carbon dioxide started to rise long before man began to produce large quantities of carbon dioxide.

Just the same it seems likely that man created carbon dioxide is not the cause of global warming, or climate change.

CHAPTER 9

Benefits of Carbon Dioxide

There is a lot of talk these days about carbon dioxide (CO_2). We hear that it is causing the temperature to rise and it will rise until it reaches a tipping point, and then the Earth as we know it will be destroyed. This is very hard to believe seeing that the Earth has been much warmer and much colder at times in the past.

But how often do we hear about the positive things concerning carbon dioxide? First off, we have to recognize that every plant and every animal on earth is made from carbon which comes from carbon dioxide: All of it. That is hard to believe since the atmosphere contains only 0.041 percent carbon dioxide.

Not only that, but plants grow better when they have more carbon dioxide. Some nurseries burn gas in the greenhouses to increase the carbon dioxide and help the plants grow.

Here is a picture of Sherwood B. Idso and pine trees at various levels of carbon dioxide. Not only do the trees grow faster with more carbon dioxide, but they are stronger and require less water. Their roots also grow faster in elevated carbon dioxide.

Carbon dioxide is a benefit to all plants and acts as a fertilizer. This is true even for truck crops or vegetables that we eat. They all grow better with more carbon dioxide.

It is very probable that our coal and oil deposits were made at a time when carbon dioxide levels were much higher than they are now.

Carbon dioxide is not pollution. Carbon dioxide levels are many times higher in submarines and on spaceships than in our atmosphere.

Carbon dioxide is also used to make dry ice (solid carbon dioxide) that is used to keep meat frozen in transit. Carbon dioxide is also used to carbonate beverages; notice the foam when you take the cap off.

Carbon dioxide is used in fire extinguishers to suppress electrical fires where using water would be dangerous. The carbon dioxide takes the place of the oxygen, and the fire cannot burn without oxygen, so it goes out.

Remember that life is a cycle: Animals take in oxygen and give off carbon dioxide. Plants take in the carbon dioxide and give off oxygen. If you were to look at a graph of carbon dioxide levels, you would see a sawtooth design: The carbon dioxide goes down in the springtime when plants begin to grow.

It is also important to point out that the oceans hold fifty times more carbon dioxide than is in the atmosphere. As the sea warms, it gives off carbon dioxide.

It is very likely that temperature is controlled by the number of sunspots on the sun rather than by the carbon dioxide produced by man.

CHAPTER 10

Capitalism Versus Environmentalism

For over fifty years, environmentalists have been primarily supported by the left. They are consistently seeking more government regulations to fix problems they honestly (or dishonestly) believe will end humanity as we know it. This is sometimes called the Chicken Little philosophy, "The sky is falling. I must go and tell the king."

All rational people are concerned about the environment and want to do all they can to protect it. But what is the best way to deal with environmental problems, more government regulations or capitalism? No matter what your position on global warming, or climate change, or the polar bears, or endangered species, or the ice caps, capitalism has a better solution than more government regulations. Capitalism seeks to protect the environment because we know that it is from the natural environment that we sustain our lives.

Most nations of the world made commitments at the Paris Conference through the Paris Accord to reduce emissions, yet they are not doing so. On the other hand, the United States, which withdrew from the Paris climate accord, is reducing emissions primarily because of switching many power plants from coal to natural gas (methane). In fact, the United States leads the world in CO_2 reduction.

Some environmentalists believe that all warming is caused by the amount of carbon dioxide in the air and that man is the primary cause of that increase in carbon dioxide. But where is the proof? Carbon dioxide levels were increasing even before the Industrial Revolution.

So our country has reduced CO_2 more than any other country in the world, and that is primarily because of pure capitalism.

Although you may not agree with President Trump on many things, you have to agree he is a capitalist and that his primary goal is to "Make America Great Again." Making America great again is simply pure capitalism.

When Trump first took office, he determined to reduce regulations, and as a result, we have seen a boon in a new technology called fracking, a process where oil is obtained from strata where it could not have been obtained before. This reduced the price of natural gas which had a huge effect on energy production, resulting in more use of natural gas to produce electricity and less coal meant less carbon dioxide.

Because land owners want to get the most they can from what they own, their care of the land has produced a reduction in carbon dioxide production. This is not only true for coal versus natural gas but in all aspects of agriculture as well. Now they are investing less and producing more food than ever before.

Now large companies are also investing less energy but producing more product than they were just a few years ago. Where the United States used to purchase most oil abroad, now we produce most of what we need ourselves.

Capitalists use less energy simply because when they save energy, they save money, and in doing so, they earn more in the end. This is capitalism. Since consumers know this, they choose producers who are more productive because their products are cheaper.

All this affects the amount of water used, which in turn protects the fish industry as well as waterfowl. Much of this has been accomplished by private industry and not by government decree. Compared to big government, people through the free market win every time. History has shown time and time again that the more government is involved in a particular industry, the more the people suffer. But with capitalism and the free market, the people are winners every time. It is evident that man is not the chief cause of warming and that minimal warming is not harmful to people.

CHAPTER 11

Carbon Dioxide

Carbon dioxide is a fairly simple compound, just one carbon atom and two atoms of oxygen CO_2. Whenever almost anything burns, the result is carbon dioxide and water. The same process takes place when things rot. A rotting tree also gives off carbon dioxide and water. So ultimately, the same thing results whether a tree burns or simply rots over several years.

Carbon dioxide is a colorless gas, so when you see pictures of smokestacks with dark vapors, it is mostly water vapor. It all depends on where the sun is. If you actually watch the smokestack long enough, you will see the vapor disappear.

You will see the same thing in the morning when you first start your car. First, a lot of water vapor, then in just a few minutes as the car warms, you do not see anything. Looking later you can even see where the water from the tailpipe dripped on the driveway.

We are told that rising carbon dioxide is causing the temperature to rise. But historically, there have been many times when the temperature fell while the carbon dioxide was still rising. In fact, there is considerable evidence that rising temperature comes before rising carbon dioxide. It is sometimes hard to tell which controls the other.

Every living thing is made of carbon in some form, every tree, every flower, every animal, and every person, even every animal in the sea. Limestone, which is calcium carbonate, is the remains of millions of sea animals. We are all made from carbon, and most of that carbon comes from the atmosphere.

It is hard to believe that when the atmospheric carbon dioxide is just 0.041 percent, it produces so many trees, plants, and animals. Carbon dioxide in the sea produces carbonate, so animals can make their shells from calcium and carbonate ($CaCO_3$). (Limestone is $CaCO_3$, calcium carbonate.) This is calcite, the mineral from which seashells are made. Limestone can also contain magnesium ($CaMg(CO_3)2$), and this is called dolomite. Pure calcite is rarely found in nature, but there are vast formations of limestone in the Rockies.

The air we breathe or our atmosphere contains the following gases: nitrogen, 78 percent; oxygen, 21 percent; and argon, 1 percent. At the present time, the atmosphere contains about 0.041 percent carbon dioxide. It also contains various amounts of water vapor depending on the location, the storm cycle, and the time of year. Some places, like Antarctica, have almost no water in the air, and it is a virtual desert.

The atmosphere also includes trace amounts of neon, helium, methane, krypton, hydrogen, nitrous oxide, xenon, ozone, iodine, carbon monoxide, and ammonia. (The atmosphere no doubt contains very small amounts of about every substance on Earth.)

As you can see, the main substances in the atmosphere are nitrogen and oxygen. Carbon dioxide is considered a trace gas, yet even at 0.041 percent, it produces all the vegetation we see on Earth. It is sometimes referred to as the gas of life.

We are told that the rising level of carbon dioxide will reach a point where we will all be dead unless trillions of dollars are spent to stop it. The truth is that the level has been much higher in the past.

Just a couple examples of toxicity: limits of carbon dioxide on submarines is 2.5 percent, and on spacecrafts, it is 1.3 percent. The difference is simply that if the level of carbon dioxide gets too high on a submarine, you can always surface, while you cannot do that on a spacecraft. The point is that 2.5 percent and 1.3 percent are many times higher than our atmospheric 0.041 percent. If people can survive on a submarine at 2.5 percent or in space at 1.3 percent, then the atmosphere on Earth would have to double many, many times before it becomes dangerous.

Some recent articles have tried to make the case that if carbon dioxide continues to rise at the present rate, then all the plants will die. However, the opposite is true. As the level of carbon dioxide has risen, the plants have grown better. Carbon dioxide acts as a plant fertilizer. Also, plants raised with higher carbon dioxide use less water, and they are more resistant to insects and disease.

Plants respire just like animals. They have tiny holes on the bottom of the leaves called "stomata," and carbon dioxide enters there, mainly in the daytime. Then through chlorophyll (which makes them green) and photosynthesis, the plant produces oxygen which it "exhales" through these same stomata. Plants also release some carbon dioxide at night but not very much. The process of life is that animals give off carbon dioxide and the plants take in the carbon dioxide and give off oxygen, so it is a very complementary process.

One of the problems with the information which has been in the press is all the negatives about carbon dioxide, but none of the many benefits of carbon dioxide.

Carbon dioxide is a very useful chemical. Frozen, it is "dry ice" used to keep frozen foods frozen in transit. Carbon dioxide is used in fire extinguishers. It is also used in various chemical and manufacturing processes.

Probably one of the most common uses of carbon dioxide is in producing the carbonation in carbonated beverages.

Carbon dioxide is also important in baking bread, and in sourdough bread, the carbon dioxide is actually produced by tiny organisms. This process takes place in many useful things.

Since all plants require carbon dioxide to grow, a number of experiments have shown that plants grow faster if grown with higher levels of carbon dioxide.

Man produces carbon dioxide by breathing and in the operation of various machinery, including our automobiles. Many efforts have been put forth to reduce carbon dioxide in automobiles, but most of them have been pretty expensive, even electric cars.

What man produces is called anthropogenic carbon dioxide. Anthropogenic simply means generated or produced by man. Some people believe that the production of carbon dioxide will rise to a point causing catastrophic global warming. They believe it could burn up earth. Yet very little of the carbon dioxide that is in the atmosphere actually comes from man. (We are told there are only twelve years left, yet just ten years ago, they said the same thing. They keep on moving the goal post.)

All their concern is based on what is called the greenhouse effect. The Earth gets warm just like your car does on a hot day. But the Earth warms and cools naturally. In some places on the Earth, the temperature may reach 100 °F in the daytime and drop below freezing at night. In winter, it gets even colder. So if there is a big difference in temperature, it is not likely that warming one degree is going to make any difference.

The effort to reduce the temperature has not been successful so far, but it is taking a lot of money. Now they are asking for a trillion dollars a year to stop global warming. Global warming is no longer a scientific theory but a political one.

This program to stop global warming is designed to take greater and greater control of our lives. This program is called collectivism. Socialism is just one type of collectivism. Many young people believe socialism is greater than our own capitalism yet have no idea what it takes to make a country go.

Rather than being a danger to all life, carbon dioxide is the greatest chemical on the Earth.

For a really interesting discussion about testing of high levels of carbon dioxide.[13]

NASA development and carbon dioxide on spacecraft: [14]

If there was no carbon dioxide, all plants and animals would die.

It seems pretty obvious that nature, not man, is the source of most carbon dioxide in the Earth's atmosphere.

---

[13] http://www.nap.edu/read/11170/chapter/5#58
[14] http://www.ntrs.nasa.gov/archive/nasa/casi.ntrs.nasa.gov/20090029352.pdf

CHAPTER 12

The Historical Perspective on Climate Change

This is an analysis of the book *Climate and Man: Yearbook of Agriculture (1941)* published by the US Department of Agriculture (1,248 pages, with index).

Introduction

1941 was a great year, among the warmest in the twentieth century. In fact, the temperature fell from about 1941 to about 1975. The perspective of this book is important because the authors were not attempting to convince people that the Earth is in grave danger because of a rising level of carbon dioxide, but these experts in their fields were working on solutions that would allow plants to live where they would not naturally grow.

This U.S. Dept of Agriculture book of 1941 needs to be compared with any pseudoscientific articles written today. Today, we have a totally different perspective. Rather than trying to help the poor of the world with clean water and good crops, we are trying to get them to use less energy, a cultural death trap. To truly understand what is happening with climate today, we need a historical perspective, and the *1941 Yearbook* gives us some of that perspective.

The book is refreshing in that it is just a matter of fact, taking life as it comes. It has many articles by many professionals, but the main value of the book is the development of life in the year 1941. Floods, hurricanes, tornadoes, fires, etc., all these are taken as a matter of fact and dealt with. The perspective is that these things have always occurred and will continue to occur without any serious consequences if proper precautions are taken. The main theme throughout the book is not how we can change the world as it is, but simply, how we can make the most of our present situation.

To read more of the analysis of this book, please see Appendix 1.

CHAPTER 13

Why Climate Change Alarmism Isn't Science

Way back in the 1970s, environmentalists were predicting global cooling. I can remember when they were talking about pumping water from Salt Lake to a valley over the hill because the rain and cold were filling the lake with water. Environmentalists predicted the end of the world because of "global cooling." Magazine covers even had pictures of a snowball Earth.

In the 1980s, the crisis ended, and they came up with a new one: global warming. This was a new catastrophe. They predicted that the poles and Greenland would melt and the sea would rise dramatically. The only solution was for the government to join the fight. Millions would die, we were told, unless the government took dramatic (and expensive) action. The only solution was for the government to act, and to act quickly.

But the Earth did not warm as predicted, so the narrative was changed to climate change, to include all dramatic weather, floods, hurricanes, forest fires, droughts, etc.

This new program is not science but pseudoscience. For you see in order for this to be science, there must be something in the theory that could be proven by experiment or observation to be false. But real science has shown us that in our long history, the Earth at various times has been warmer as well as cooler. Further, the estimates that were developed never came to pass. One even predicted the end of the Earth in just ten years unless we spent vast quantities of money; but that was over twelve years ago.

But just because the climate changes does not prove anything, as the climate has changed from year to year, from decade to decade, and from century to century.

They said that they had proven their theory, but no science is ever proven true. No matter how much effort may demonstrate it to be true, one experiment or observation may prove it to be false.

Climate "science" today is based on the one-sided notion that man is the only cause of global warming and the effort is concentrated on reducing emissions. But no effort is put forth to learn anything about nature, the deep oceans, other greenhouse gases such as water vapor, and the sun, the originator of all energy on the Earth, or even how much heat comes from deep in the Earth. (If you go deep in a mine, you will find it is very hot unless there is forced ventilation.) The challenge of the Intergovernmental Panel on Climate Change (IPCC) was not an open-ended study of climate, but simply that man was the cause and what could be done about it. How, then, would they find out what part nature plays in climate change? Unless you look for something, you will never find it.

Therefore, instead of trying to find the solutions to global warming and mitigate anything that may need it, all they seem to be able to do is make doomsday predictions.

Therefore, when you look at their predictions, even by great men like Dr. James Hansen and Al Gore, none of them have come true in the time predicted. Some of their predictions were very specific, such as wind damage and sea level rise. One person recently gave the Earth only fifty days survival.

They tell us that to solve the problem of climate change, it is going to take billions (even trillions) of dollars, and each one of us is going to have to suffer to make it happen.

But climate change is clearly a scam. It is based on fear, our natural fear of calamity. Most of us have experienced a flood, a fire, or an earthquake and know what these things mean. These radicals depend more on our fear and the need to do at least something to stop this so-called calamity. It never seems to occur to any of them that the real solution is to just do nothing. Nature has a way of seeking a balance.

They say there is a consensus and that 97 percent of scientists agree. Yet science is not based on consensus but on proven theories. (I had a teacher who did not believe the "law" of gravity had been proven and said that when he dropped a ball, he always looked up so if it happened to go up rather than down; he wanted to be sure to see it.)

When we look at the history of the Earth, we find long periods of time when it was a lot warmer than it is now and other times when it was a lot colder. Where we live in North Idaho at one time, ice was a mile deep, and how about the ice that was in Yosemite National Park, the perfect glacial valley? At one time, the Thames River froze over for many years, and there were other times when they grew grapes in the far north country of Europe.

These times are called the Little Ice Age, which ended in about 1850, and the Medieval Warm Period, which was before that.

There has never been a debate between the climate alarmists and the skeptics. Notice they are not deniers, like those who deny the Holocaust or the moon landing. For the alarmists are so sure they are right, and they are so busy trying to save the world, they just do not have time to discuss anything. It is really too bad. If they know they are right, what have they to fear?

Man is not the cause of global warming, and climate change is a scam.

(Based on an article by Daniel G. Jones)

CHAPTER 14

Climate Change Settled

Climate science has only been of any importance for about the last thirty (30) years. People have always been interested in knowing the weather, but there were few if any climate scientists and the few that there were, were just learning the trade.

Some scientists have decided that the Earth as we know it is going to be destroyed unless we take immediate action to stop the use of fossil fuels and convert everything to wind and solar. The prediction was made over twelve years ago that we only had about ten (10) years left. Now they are predicting again, and this time it is twelve (12) years.

The difficulty is that few of these "great" scientists want to debate the issue because the science, they say, is settled and they are too busy "saving the world" to spend any time in debate. It is possible that they do not want to debate because they are not absolutely sure they are right. We are told that 97 percent of scientists agree that the science is settled and the world as we know it will end unless we take immediate action. And that action will be very expensive and will require many more layers of government. (They were even upset when a climate skeptic was to be part of a climate committee.)

The 97 percent turns out is only 0.3 percent when you actually get down to what these scientists were saying they actually agreed to.

Science does not operate on consensus. You have a theory and you set out to prove the theory. You actually have two theories, one negative and one positive. Then you collect data. If the data is positive, you put it on the positive shelf, and if negative, you put it on the negative shelf. This is what is called a hypothesis. You can make up your mind and then do your research to prove you are right, but that is not science. History is full of folk who everyone said was wrong but was later proved to be right. There was actually a time when some even thought the Earth was flat or the sun went around earth.

The point is that if you are sure you are right, you should not be afraid to debate with others because the data might demonstrate that you are wrong. But to cut off all debate, as has been done in some circles lately, is completely unscientific. True scientists must be willing to test their hypothesis to see if it is actually right.

As has been pointed out, it is harder to prove yourself wrong if doing so will mean that all the government money you have been getting will be cut off.

There is so much about our present ideas that are simply not consistent with our scientific history. Our country has become great because of the use of fossil fuels, and we ought not to stop their use until, and unless, we can demonstrate that they are no longer necessary. Renewables only account for a very small percentage of our power, and if we depend on them totally, we may find ourselves living in the dark. This is a serious problem, and we need to be careful not to leap before we look.

No, the science is not settled, and we need many more who will test the so-called science to see if it is true. Once we have done that, maybe we will have a consensus, like we have with the law of gravity. The simple truth of the matter is that the science is not settled.

Global warming has no doubt brought many to work in the field simply because they have some sort of government assistance, not because what they know is valid.

CHAPTER 15

COP 24 Climate Conferences

COP stands for Conference of the Parties, and one conference is held each year. These conferences have been held every year starting in 1995. Here is a list of the conferences and where they were conducted.

COP 1	March 28 to April 17	1995	Berlin, Germany
COP 2	July 8 to July 19	1996	Geneva, Switzerland
COP 3	Dec 1 to Dec 10	1997	Kyoto, Japan; Kyoto Protocol; failed
COP 4	Nov 2 to Nov 13	1998	Buenos Aires, Argentina
COP 5	Oct 25 to Nov 5	1999	Bonn, Germany
COP 6	Nov 13 to Nov 2	2000	The Hague, Netherlands
COP 7	Oct 29 to Nov 10	2001	Marrakech, Morocco
COP 8	Oct 23 to Nov 1	2002	New Delhi, India
COP 9	Dec 1 to Dec 12	2003	Milan, Italy
COP 10	Dec 6 to Dec 17	2004	Buenos Aires, Argentina
COP 11	Nov 28 to Dec 9	2005	Montreal, Canada
COP 12	Nov 6 to Nov17	2006	Nairobi, Kenya
COP 13	Dec 3 to Dec 17	2007	Bali, Indonesia
COP 14	Dec 1 to Dec 12	2008	Poznan, Poland
COP 15	Dec 7 to Dec 18	2009	Copenhagen, Denmark
COP 16	Nov 28 to Dec 10	2010	Cancun, Mexico
COP 17	Nov 28 to Dec 9	2011	Durban, South Africa

COP 18	Nov 26 to Dec 7	2012	Doha, Qatar
COP 19	Nov 11 to Nov 23	2013	Warsaw, Poland
COP 20	Dec 1 to Dec 12	2014	Lima, Peru
COP 21	Nov 30 to Dec 12	2015	Paris, France (Paris Agreement)
COP 22	Nov 7 to Nov 18	2016	Marrakech, Morocco
COP 23	Nov 6 to Nov 17	2017	Bonn, Germany
COP 24	Dec 3 to Dec 14	2018	Katowice, Poland
COP 25	Nov 11 to Nov 22	2019	It was to be in Brazil. They withdrew. Now Chile

Was Poland's COP 24 climate change conference the end?

Dennis Avery, Climatologist, has some interesting comments about COP 24:

Posted: Dec 10, 2018 11:32 AM

The opinions expressed by columnists are their own and do not represent the views of Townhall.com.

I'm just back from Katowice, Poland—site of the UN Intergovernmental Panel on Climate Change's latest climate change conference. The UN had been threatening to inflict such meetings on the world every year until we surrendered to their carbon taxes. Today, however, I wonder if Katowice—the 24th—will turn out to be the last such IPCC climate get-together?

The global outlook for Green Energy has suddenly gone from weak to bleak as international revolts against carbon taxes go viral.

In France, President Macron retreated from view as truck drivers, farmers, and students joined the Yellow Vest movement in the fourth straight weekend of riots. They're protesting Macron's promises to keep raising French fuel taxes ever-higher—to force coal, gasoline, and fuel oil out of French daily life. (Without France's many nuclear plants—which produce *no* CO2—the electric costs in France would be vastly higher.)

Further broadening the French protests, a major police union also called for an 'unlimited strike' in solidarity with the Yellow Vests. The union complained that the French government wanted the police to "take the blows" instead of its policymakers.

The Washington Post said, "The sight of one of Europe's most climate-ambitious countries beating a hasty retreat over a proposal that would have hiked gasoline tax by 4 cents, or just under 3 percent, high-lighted the difficulty of imposing any economic pain in the name of tackling climate change." But the *Post* is being disingenuous. The 4-cent tax hike was clearly billed as just the first step in Macron's "whatever it takes to kill fossil fuels" taxation regime.

In Germany, the fabulous electrical cost overruns from Germany's solar panels and wind turbines are driving long-time Chancellor Angela Merkel out of office. However, she is trying to leave the ever-costlier renewables policy in place, and even leaving her mandate to close Germany's remaining nuclear plants. (That was issued after the tsunami hit Japan's Fukushima nuclear plants in 2011, though tsunamis are unknown on Germany's Baltic seacoast.)

Germany's powerful chemical and auto companies now fear they'll have to build their future plants in lower-cost energy markets such as China, India, and the United States.

In Ontario, Canada, Premier Kathleen Wynne's government was dumped due to the nationwide carbon tax plan. New Premier Doug

Ford and his conservative coalition promised to reject carbon taxes. Australian lawmakers voted to repeal their country's carbon tax in 2014.

In Katowice, my Heartland Institute team was approached Wednesday by Solidarity, the Polish ship-builders union that brought down Poland's Communist Government in the 1980s. Solidarity immediately signed a working agreement with Heartland, aimed at helping protect Poland's continued use of coal. The EU has threatened massive fines if the Poles keep mining the coal that is their only major energy source.

In Katowice, Heartland offered speakers who explained that the sun, not CO_2, was the real source of the Modern Warming. They noted that the Earth's temperatures have trended upward since the depths of the Little Ice Age. In 1715, the new telescopes saw a weak sun with only 1.5 sunspots per day. Three hundred years later, in 2001, a much stronger sun created 170 sunspots per day. The correlation between atmospheric CO_2 and the Earth's temperature over the last century is a weak 19 percent. The correlation between sunspots and the Earth's temperatures is a powerful 79 percent.

The world's top particle physics laboratory, the Geneva-based CERN, found in 2016 that the Modern Warming is just the next phase of the long, natural Dansgaard-Oeschger Cycle. That cycle since 600 AD has brought us successively the Dark Ages, the Medieval Warming, the Little Ice Age and the Modern Warming.

A recent study of cornfields in Illinois over the past 60 years found *no increase* in field temperatures.(Study of cornfield temperatures.) In the locations where there was significant temperature change, it was to *lower* temperatures rather than higher. What does that tell us about the litany of "hottest year ever" claims from NOAA and NASA?

Corn does best when the ground has reached a certain temperature, not the air.[15]

At this point, only a minority of Americans believe they are seriously threatened by man-made warming. They are unwilling to give the U.S. government the power that Germany gave to its government—to triple and quadruple their energy costs. Four-dollar gasoline is still too fresh a memory. If Katowice, Poland, turns out to be the last UN IPCC climate conference, they're likely to applaud.

There was just nothing in the recent Poland conference to prove that man is the chief cause of global warming and that warming will lead to disaster.

---

[15] http://www.crops.extension.iastate.edu/cropnews/2012/03/influence-soil-temperature-corn-germination-and-growth

CHAPTER 16

Coral Reefs Dying

Widespread warming of coral reefs causes bleaching. This is shown to be a natural response to temperature changes of water and is a purely natural happening. Coral has warmed and cooled for thousands of years. The resulting change in the appearance of corals around the world was enough for many to proclaim that the world's corals were dying off en masse due to global warming, despite the fact that corals have lived through hundreds of relatively recent historical cycles of glacial periods interspersed with warm periods, as well as higher acid concentrations in the water and sea level rise of as much as several hundred feet.

The change in color of the coral is because of a rejection of the resident algae. When a new algae moves in, it changes the color of the coral, so that one might think it had died. The algae not only provides the coral its color but the food it needs to live.

Corals have lived here for several thousand years, so during that time the Earth must have seen considerable warming and cooling. Bleaching is just a normal part of its natural cycle.

Bleaching corals is not a sign that the world is about to end because of global warming but just a normal part of coral's natural life cycle.

The bleaching of corals is not a sign that climate change is killing them or that man's use of fossil fuels is bringing the Earth to a dramatic end. And it seems evident that man is not the cause of any bleaching of corals.

CHAPTER 17

DDT Pesticide: *Silent Spring*

In 1962, Rachel Carson wrote a book called *Silent Spring*, in which she insisted that birds and fish were dying because of DDT. However, DDT has been successfully used in Africa and Europe to kill insects that carried serious plagues. Nonetheless, DDT was ultimately nearly banned with the result that many natives, especially in Africa, died of malaria.

It is also possible that DDT did not cause the death of large numbers of birds.

The publication of *Silent Spring*, which led to the partial ban of DDT, was very controversial at the time and even fifty years later. Dichlorodiphenyltrichloroethane (DDT) is one of the best known of all the synthetic insecticides. Rather than just banning it, more effort should have been taken to determine why it was effective in some situations and not in others. Maybe a program could be developed which would save the birds while also protecting the lives of those who lived around infectious mosquitoes.

Of immediate concern to the military during World War II was the possible use of DDT for the control of several insect-borne diseases: malaria (carried by *Anopheles* mosquitoes), typhus (carried by body lice), and dysentery and typhoid fever (carried by houseflies). They had been searching for a substitute for pyrethrum made from chrysanthemum flowers imported chiefly from Japan. But the war with Japan had cut off the source of supply just as the demand for pyrethrum soared.

Once DDT was determined effective against these disease-carrying insects, it was quickly put into large-scale production. This seemed like a real answer because it was easy to produce and safe to handle. Trials helped overcome small typhus epidemics in Mexico, Algeria, and Egypt. Soon, the process was used in Naples where powdered DDT was blown down the clothing. This stopped a potential typhus epidemic. Before the treatment, as many as sixty cases of typhus were contracted a day, and people were dying by the score. But after dusting, they were able to see new cases decline rapidly.

Similar results were found against malaria and the *Anopheles* mosquitoes.

The DDT was sprayed on walls, and it lasted over six months. Millions of soldiers even carried small spray cans, and aerosol bombs were used in the tents and buildings.

Malaria had been declining in the United States because of screens and draining of swamps.

After the war, the US Public Health Service began a large-scale program of malaria eradication using DDT. Large-scale eradication programs were also undertaken in Africa and India. After DDT was introduced in Ceylon (or Sri Lanka), the number of malaria cases fell from 2.8 million in 1946 to just 110 in 1961. Same with Formosa (Taiwan) which went from over one million cases to only nine. Similar decreases of malaria cases and deaths were seen everywhere DDT was used.

Although DDT was so successful against diseases carried by insects, people were cautioned to keep it away from foods.

Since 90 percent of insects do something good, it was important to find some way to make a distinction. Robins that ate earthworms that had eaten leaves that had been disinfected died. So not everything was rosy with DDT.

The importance of Rachel Carson and *Silent Spring* is that it fathered the environmental movement with all the positives and negatives associated with it.

DDT has not actually been banned, but its use has been very specific and short-termed. DDT is neither a panacea nor a super villain. In many places, DDT failed to eradicate malaria, not because of restrictions on its use, but because it simply stopped working, it was no longer effective. Insects developed immunity to the pesticide.

It is also evident that no amount of elimination of DDT will affect man caused global warming.

CHAPTER 18

Dr. Willie Soon Is Right

And the global warming apocalypse is not nigh.
Real-world evidence certainly supports him.
—Jeffrey Foss, PhD

(The following is attributed to Dr. Jeffrey Foss.)

Everyone has heard the bad news. Imminent climate apocalypse (aka "global warming" and "climate change") threatens humanity and the planet with devastation, unless we abandon the use of fossil fuels.

Far fewer people have heard the good news. The sun has just entered its grand minimum phase, and the Earth will gradually cool over the next few decades.

Why should we all hope the Earth will cool? Because nobody with any trace of human decency would *hope* the Earth will actually suffer catastrophic warming. (Of course, global cooling could cause serious crop failures, which would be just as devastating.)

Many of us *believe* in the threat of global warming but live in the hope that we can switch to "renewable energy" before it is too late. But this is a false hope. Despite our best efforts over several decades, renewables such as wind and solar energy still meet only 2 percent of global energy needs, while hydro adds only 7 percent or so.

Therefore, avoiding the alleged climate apocalypse by relying on renewable energy would require surviving on less than 10 percent of our current energy requirements. But that is impossible. It would also be really catastrophic: billions could die.

Our global economy runs on energy, and over 80 percent of it is still fossil fuels, with nuclear and other nonrenewables providing another 10 percent. If we switch to renewables tomorrow, 90 percent of our energy will be lost, and the global economy will sink like the Titanic. Keeping nuclear power would merely add a second lifeboat as the great ship sinks. Even if the energy loss were spread out over decades, the final result would still be the same.

Humankind could not produce enough food, clothing, and shelter. Jobs would vanish. Massive starvation, disease, and death would result. Hard physical labor would once again become the norm. Even though life could be maintained for some portion of humanity, liberty and happiness would be lost.

Let's stop pretending. The prescribed cure for climate/global warming apocalypse is far worse than the purported disease. If we don't use coal, oil, and natural gas for energy, many of the seven billion of us now alive must die. Those who survive will be impoverished and enslaved, toiling and scavenging for food by day, and fearing the darkness by night—except for the privileged few who still have money, energy, and power.

The sudden and dramatic growth of human life, liberty, and happiness since the Industrial Revolution was achieved by replacing muscle power with coal and oil power. Before that, Hillsdale College professor of history Burt Folsom points out only the wealthy could afford whale oil and candles. Everyone else had to go to bed early, and often hungry, when the sun went down, sleeping to recover enough energy to work—only to repeat the daily cycle yet again. Freedom of thought and travel had little real worth when we were too tired to think or walk.

The petroleum age saved whales from the brink of extinction and brought cheap kerosene to the masses, so that they could read at night, bringing light into their lives and their brains.

The premature switch to renewable energy recommended by the false prophets of climate apocalypse is really just one step in an industrial counterrevolution devoutly desired by those discontented with modern life in our free market republic and other world democracies and ready to erase our hard-won prosperity and freedom.

The climate apocalypse global warming bad news is rewarded by big money from the government and servile amplification from traditional big news media, while the good news of global cooling is silenced and unheard, stifled by both traditional media and most of today's social media platforms.

We should all be suspicious of the motives of those who push this bad news and welcome those who push back. Dr. Willie Soon is one scientist, although by no means the only one, who has the courage to stand up to big money, big government, big (pseudo-) science, big media, and big environmentalism to spread the good news. It's high time we all heard it.

The good news from Dr. Soon and his fellow solar scientists is that the increase in global temperatures since 1800 was caused by two centuries of increasing solar output, not by human use of coal and oil.

But then solar output began to fall around 2000, in a repetition of a well-known two-hundred-year cycle of solar activity, and global warming stopped. That's more good news that too few people know. The purveyors of climate apocalypse have no explanation for this two-decade failure of their prophecy, which fortunately for all of humanity shows the superiority of solar science over apocalyptic warming foretold by computer models, hysteria, and headlines but not by real-world evidence.

Finally, solar science says we should expect steady but manageable global cooling until about mid-century, when solar activity will recover and temperatures begin to warm once again. Once again, this will be due to solar activity and not to fossil fuels or carbon dioxide emissions.

And the best news of all, humanity's successful pursuit of life, liberty, happiness, and better living standards and health care needn't be stopped by climate apocalypse or its prescribed cure. The only thing we have to fear is the fear of climate apocalypse itself.

Equally important, a warmer *or* cooler planet with more atmospheric CO_2 and plentiful, reliable, affordable fossil fuel and nuclear energy would be infinitely preferable to a cooler planet with less CO_2 and only expensive, intermittent, weather-dependent wind, solar, and biofuel energy.

At the very least, humankind has an historic opportunity to witness a crucial test between two scientific hypotheses of enormous consequence. The next decade or two will reveal whether the Earth warms or cools.

Surely, all right-minded people must hope that it cools and that the fear-mongering of imminent global warming apocalypse cools as well.

I might add that no one should wish the current severe Chicago-style polar vortex cold on anyone. I extend my sympathies and prayers to all who are now suffering from the cold. But be of good cheer in the knowledge that this cold snap at least puts the lie to vastly worse climate scare global warming stories.

I also wouldn't wish on anyone the "Green New Deal" energy reality of February 1, 2019, when wind power provided 15 percent of the energy that kept lights on and homes warm in America's mid-Atlantic region, solar provided *zero*, and derided and despised coal, natural gas, and nuclear power provided a whopping 93 percent of that energy! Imagine the cold, misery, and death toll under 100 percent pseudo-renewable energy.

Dr. Jeffrey Foss is a philosopher of science, professor emeritus at the University of Victoria, Canada, and author of *Beyond Environmentalism: A Philosophy of Nature.* (Other sources suggest 6.6% for wind and 93.4% for all the rest, coal, hydro, nuclear misc etc)

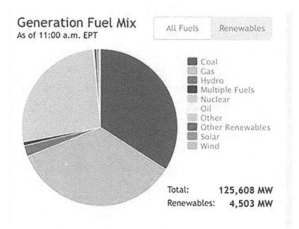

Source: PJM Interconnection regional electricity
transmitter (used with permission)

Just a short comment: I think it would be best if we wished for a gentle warming as being the best solution for our crops, many of which right now have a very short growing season. Nonetheless, it is evident that man is not the cause of the so-called global warming.

CHAPTER 19

Droughts And Rainfall

Droughts are just part of the natural variation in weather and climate. Just about the time you think it will never rain, then it rains and fills all the reservoirs. Of course, there are always going to be some droughts that are caused by man, through irrigation policies, etc.

Droughts are common, especially in the more temperate parts of the Earth. As the seasons vary, the climate changes. It brings the spring rain and the winter snows. Areas that have seen drought for several years may see several years with plenty of rain. As the climate changes, it can lead to less rain. It can also lead to more rain.

The Southwest United States saw several years of drought, then years of rain. The Midwest had several years of drought and dust bowls which caused many families to pack up and move out west. Within the last decade, drought conditions have hit the Southeastern United States, the Midwest, and the Western United States. In 2011, Texas had the driest year since 1895. In 2013, California had the driest year on record.

Probably farmers are most concerned about drought, especially those who dry farm, which is farming without irrigation. This is especially common in the Midwest, but there are many places in the United States that depend on rainwater for their crops to survive.

Climate change can affect drought, but it is important to keep in mind that climate is always changing and can cause drought one year but not the next. Some of the most interesting stories from our nation's past have been written around how families survived drought conditions.

Areas that depend on the mountain snowpack are especially susceptible to drought. When there are years with little snow, there is little runoff, and this can lead to drought.

It is more important that precipitation falls as snow than as rain, as the snow melts slowly and this affects the water in the ground.

One of the most dramatic droughts concerned the Aral Sea. Lake Aral, or the Aral Sea, at one time was the fourth largest body of water in the world. Now it has lost over 90 percent of its area. At one time, it had a vital fishing industry. Now, no more. Fishing boats are left high and dry. In 1960, Russia diverted the rivers, which fed the lake, for irrigation projects, which caused the lake to dry up. The lake originally had 26,300 square miles of area, but since 1960, it has been continually shrinking. This lake has no outlet.

Many droughts in the Southwest were overcome by diverting water from the Colorado River, and rivers from the high sierra, that fed lakes like Mono Lake. Much of this water went to provide water for the city of Los Angeles.

Droughts can be avoided by the use of good planning: Mainly dams hold winter water for use in irrigation. Unfortunately, some efforts have been made to save certain fish, which means less water for orchards and ranches. Maybe they will think of a better way to save the fish and provide water for California ranches.

While global warming and climate change can affect drought, it is evident that droughts existed long before man lived in the United States and has had little effect on the climate.

CHAPTER 20

Electric Cars in Cold and Hot Weather

The effects of hot and cold weather on battery-storage electric autos is easily defined.

The Tesla has an 85 kWh battery (depending on model).

Hot and often humid weather, above 29°C [84°F], is easily ignored in a well air-conditioned auto on a long trip or in commuter "stop and go." The numbers slightly vary from auto to auto, but the auto air-conditioner is typically 5 kWh in output. The large value is due to the movement of an auto with limited insulation in high insolation (from the sun) or in conditions of both high temperature *and* humidity, and the need to provide essential ventilation, and all this happens at speeds from 50 to 120 kph (30–72 mph).

Naturally, northern Europeans will hardly understand AC at all, but they understand *heating*.

Heating requires a similar energy to cooling, for the same but inverted reasons, 3–5 kWh. In an auto with an internal combustion engine, heating is not a problem since the waste heat from the engine will be 15 kWh or more. Provision must be made to exchange the heat and that is standard. There is a radiator for the engine and a similar radiator in a heater for the interior of the automobile.

In a battery-storage electric auto, with an electric motor in each wheel, there is not so much waste heat to heat the auto (the waste heat is back at the power plant providing the electricity to *charge* the battery in the first place), and thus, it *must* be taken from the battery.

So *hot* or *cold*, a battery is at a disadvantage.

When it is hot, the battery must supply 5 kWh of cooling for the duration of the travel, hopefully *not* a three-hour "stuck-in-hot-traffic" commute but a road trip of three hours or longer.

That means 3–5 kWh of the 85 kWh goes to cooling and reduces the 200-mile range to 160 miles.

There are other battery losses: battery losses: driving through hills, the need to accelerate, or simply to drive over 120 kph (72 mph) which increases the energy required as the *cube* of the velocity.

But that is aside from the heating and cooling.

On the heating side, even in semitropical California, heating is required in winter or in the mountains or high desert. In North Dakota, a small volcano would be welcomed inside an auto.

For a gasoline or diesel engine, the interior heating is taken almost entirely from the waste heat. In a battery-storage auto, it comes from the battery since the electric motors generate little waste heat, and it would be a problem to route this limited heat output to the interior of the auto. So a *real* heater is required to provide up to 5 kWh of heating.

The battery also must be heated to operational temperature between 59 and 77°F before drawing current. That can be done in your garage if you are at home, but otherwise, that too comes from the stored energy. The result is much the same as in the summer; the driving range is reduced by about 20 percent. Many Tesla owners complain of this issue on the Tesla user sites.

It is also a fact that the battery does not enjoy complete charge or discharge, running best between 20 percent on the low side to 85 percent on the high side; therefore, all the range numbers are smaller than shown.

In my opinion, these are all short-term issues since the battery storage in an auto is a stopgap and will become the Edsel of the near-term future, and [Elon] Musk [architect of Tesla electric cars] will still be a billionaire.

(William H. Smith, February 7, 2019; used with permission)

One thing to think about is the amount of energy it will take to melt the frost on the windows of an electric car. On a cold morning here in Idaho, it takes usually about fifteen minutes with the motor running to melt the ice from the front and rear windows. This use of energy will of necessity reduce the length of a trip on a charge for an electric vehicle. Even if man were the cause of global warming, electric cars are only a short-term solution, not usable on long trips.

CHAPTER 21

Electric Cars: A History

Introduction

An electric car is a plug-in electric automobile that is propelled by one or more electric motors, using energy typically stored in rechargeable batteries. Since 2008, a renaissance in electric vehicle manufacturing occurred due to advances in batteries, concerns about increasing oil prices, and the desire to reduce greenhouse gas emissions. Several national and local governments have established tax credits, subsidies, and other incentives to promote the introduction and adoption in the mass manufacture of electric vehicles.

Early electric car history

The first electric car in the United States was developed in 1890–1891 by William Morrison of Des Moines, Iowa; the vehicle was a six-passenger wagon capable of reaching a speed of 23 kph (14 mph). Electric cars did not catch on until the average home was wired for electricity. By the turn of the century, many homes were wired for electricity such that 22 percent of cars were driven by gasoline, 40 percent by steam, and 38 percent by electricity. These cars were massive and sold to more wealthy customers. They had luxurious interiors and used considerable expensive materials. The sales of electric cars peaked in the early 1910s.

By the turn of the century, electric cars began to lose out in the market primarily because of the discovery of oil and the invention of the internal combustion engine, which was more practical on long trips. The electric cars were limited for use in the cities because their speed was no more than 15–20 mph, and their trip length was limited to 30–40 miles. Gasoline cars were now able to travel farther and faster than the equivalent electric ones.

Automobiles with internal combustion engines that burned gasoline were even better because of the invention of the electric starter which replaced the hand crank. Then Henry Ford perfected mass production, and this enabled the production of cars at a much lower price.

In the 1960s, all electric vehicles were designed so that they could be easily charged at charging stations. These were considered plug-in models.

The vehicles used on the moon were called Lunar Roving Vehicles and were battery-powered. Three of those vehicles are still on the moon. (The Russians have two of them up there.)

Energy crisis helps provide incentive for electric cars

The energy crises of the 1970s and 1980s brought renewed interest in electric cars. However, during the 1990s, the interest turned to gasoline-powered sport utility vehicles as being more practical than even other gasoline- or diesel-powered vehicles. There is still a great market for these vehicles.

Also, along about this time, diesel-powered vehicles, especially pickups, became popular because at that time diesel was a lot cheaper than gasoline. And diesel vehicles, though somewhat more expensive, had a longer life.

Although the electric vehicles produced today are far better than the original ones, they are still limited to short trips and slow speeds. Even with that, thousands of these vehicles have been sold.

First all-electric vehicle

A California car maker, Tesla Motors, began producing the Tesla Roadster in 2004, which was first delivered to customers in 2008. It was the first all-electric vehicle to travel 200 miles on a charge. Other car manufacturers have produced a number of new models which are still being sold.

There is a present move to require everyone to own all electric vehicles, but little attention has been given to how these are powered, even with multiple charge stations. So where does the power come from that will charge these automobiles? Some will be powered by solar and wind, some by hydroelectric, some by nuclear, and some by gas, but if they are powered with coal, then they have gained nothing in terms of reducing any greenhouse global warming but will have spent lots of money on newer vehicles. With any new idea, care must be given to understand all the ramifications associated with the product. [16]

Electrical cars today have a lot of problems. They are slow in speed and limited in miles they can cover. The batteries wear out and must be replaced and are fairly expensive.

One present proposal includes many charging stations all over the country, but no information is given on who is going to provide the charging stations or who is going to pay for the electricity used by each one. Using electric cars is not going to solve any supposed global warming problem, especially any part that man has in producing carbon dioxide.

Here is another problem: warming the car in cold, or even very cold weather, and cooling the car in summer (see "Electric Cars in Cold and Hot Weather").

---

[16] http://www.en.wikipedia.org/wiki/History_of_the_electric_vehicle

CHAPTER 22

Extinctions: Animal and Plant

Flightless dodo bird

Several authors have talked about species extinctions unless we stop the use of fossil fuels. However, the examples they give are not very truthful.

The dodos were found on the island of Mauritius, and Dutch sailors killed them and ate them. They were a fairly large bird that did not fly, so easy to catch.

The passenger pigeons used to black out the sun when they flew over in the West, yet with the new railroad, they were caught and sent by rail to restaurants back east as a delicacy. Eventually, there were none left.

Other birds became extinct because of a change in their habitat, not because of global warming.

Buffalo became almost extinct because they were killed for their hides. Native Americans drove them off cliffs to kill them for their hides. Fortunately, several people rescued them, and now, there are even buffalo ranches, and in some locations, you can buy buffalo hamburger.

We were told that plants would become extinct because their seeds could not pass through paved cities. But those who make such statements have never planted a garden. Native plants spring up in the spring because the wind carries the seeds, or birds eat them and drop the seeds. Even if they did become extinct because of cities, it would not be because of global warming.

Most animals range over a large area and pick the place that suits them the most. Even polar bears and grizzlies sometimes manage to get together, and their habitats are far apart.

The change in temperature of just a couple of degree will not affect any plants or animals. Scientists have discovered that with warmer weather, plants move farther north and higher up on the mountains but do not abandon their original habitat.

Warmer weather is better for most animals and most plants. You can buy plants at the nursery which will not survive in the north: They don't grow because it is too cold, not because it is too warm.

Thousands of species have become extinct in the past but not today. And even if it should happen, it would not be because of global warming. More importantly, man is not the cause of any species extinction because of the so-called global warming.

CHAPTER 23

Extreme Weather

Probably the most important flood event was the Johnstown Flood. The earthen dam was not properly maintained, and the residents were seemingly immune to disaster. On May 31, 1889, the dam, holding back a large lake, burst and took with it every small town on the way to Johnstown and then Johnstown itself. In the end, 2,209 people died. Those who had not drowned died in the fire that was ignited at the stone bridge where many buildings were piled up.

Thousands of people died simply because they had no way of knowing that a flood was coming their way. A telegraph message was sent to Johnstown but did not reach all the people.[17]

There were other great disasters. One was along the Florida Keys. The wind and rain washed out a new railroad which was never rebuilt.

There are as many examples of extreme weather hundreds of years ago as there are today. The question, however, is this; are things worse today than they were back in those days? There is considerable evidence that things have been worse in the past. The difficulty today is that many more expensive buildings are often in the path of great storms.

---

[17] https://www.youtube.com/watch?v=Q62vlcFLLlM
(For more information, see the chapter "Floods".)

The question has been asked if climate change or global warming is making extreme weather worse. It would be hard to prove, especially since historically there have been many serious climate events.

One writer put it this way:

> Debates about whether single events are "caused" by climate change are illogical, but individual events offer important lessons about society's vulnerabilities to climate change. Reducing the future risk of extreme weather requires reducing greenhouse gas emissions and adapting to changes that are already unavoidable. ("White Paper: Center For Climate and Energy Solutions, 2011–2012")

Looking at the historical record, it is evident that, while things seem more dramatic today, there have been many very dramatic weather events, even in the recent past. Going back even further, we can see some very dramatic events, like when the Missoula ice dam broke and water ran over the West. This flood produced the Channeled Scablands found in some of the northwest states.

One thing seems pretty clear, and that is that man is not the author of global warming that will destroy the Earth in as little as ten years.

CHAPTER 24

Floods

The Johnstown Flood in a river valley in central Pennsylvania happened on May 31, 1889. The South Fork Dam on the Little Conemaugh River was owned by a resort and for many years had little maintenance. So when there was an unusual rainstorm, the water ran over the top of the dam and washed the dam out; as much water ran down the valley as runs in the Mississippi River.

Someone telegraphed to Johnstown, but for some reason, the word did not get to all those concerned.

There was a stone bridge down river, and many of the floating houses were held back by the bridge. Then they caught on fire, and many of the people who were not drowned were killed in the fire. There was $17 million in damages, and 2,209 people lost their lives.

The American Red Cross, led by Clara Barton, and fifty volunteers undertook a major disaster relief effort. Support for the victims came from all over the United States and eighteen foreign countries. The victims sued the club but were never able to recover damages.

Here are some facts about the flood:

1. 2,209 people died.
2. Nine entire families died, including 396 children.
3. 124 women and 198 men lost their spouses.
4. More than 777 victims were never identified and rest in the Plot of the Unknown in Grandview Cemetery.

5. Bodies were found as far away as Cincinnati and as late as 1911.
6. 1,600 homes were destroyed.
7. $17 million in property damage was done.
8. Four square miles of downtown Johnstown were completely destroyed.
9. The pile of debris at the stone bridge covered thirty acres.
10. The distance between the dam and Johnstown was fourteen miles.
11. The dam was owned by the South Fork Fishing and Hunting Club, an exclusive club that counted Andrew Carnegie and Henry Clay Frick among its members.
12. Flood lines were found as high as eighty-nine feet above the river level.
13. The wave from the flood measured thirty-five to forty feet high and hit Johnstown at forty miles per hour.
14. The force of the flood swept several locomotives weighing 170,000 pounds as far as 4,800 ft.
15. $3,742,818.78 was collected for the Johnstown relief effort from the United States and eighteen foreign countries.
16. The American Red Cross, led by Clara Barton and organized in 1881, arrived in Johnstown on June 5, 1889; this was the first major peacetime disaster relief effort by the Red Cross.
17. Although the Johnstown Flood was without doubt the most dramatic and senseless flood in the history of the United States, there were also other floods caused by hurricanes, such as the Galveston flood, September 8, 1900, in which there was a 15-ft storm surge, and eight thousand people died. There were also other hurricanes which flooded the Florida Keys and other parts of the South. One destroyed the railway on the Keys, which was never rebuilt.

Some people have connected flooding with climate change. For one thing, it is very hard to imagine a one-degree increase in temperature having a dramatic effect on flooding. Here are just a few things to keep in mind:

1. A warmer atmosphere holds more moisture. But the climate change link to flooding is very complicated, and climate change may not mean more and heavier rains.

2. Evidence of heavier rainfall in the past is limited, but knowledge is growing. It is hard to distinguish between what is a climate cataclysm and just normal weather variability. For one thing, we have not been keeping records long enough to be able to separate the one from the other.

3. Attributing specific events to climate change is tricky, and flooding is no exception. If there is doubt, we should consider this just normal variation of the weather cycle.

4. Scientists predict that heavy rainfall will increase in the future. (I remember when we lived in Pensacola, Florida, that it would rain like crazy, but in just a few minutes, the water would all be gone.)

5. Flooding isn't just about rainfall; other human factors contribute too. Paving of large areas, building many more homes, and closing of waterways are only a few of the factors that affect flooding. The more impervious areas there are, the more the area is affected by flooding.

6. Of course, rapidly melting snow can be a big factor in flooding.

Another great flood was the Houston flood caused by Hurricane Harvey, August 12, 2017. Water was measured from fifty to sixty inches of rain. The biggest reason was that the hurricane simply parked over Houston, and it kept on raining. Strong, heavy rains that stay in one place can do a lot of damage.

2005 saw the flooding in New Orleans from Hurricane Katrina. Much of the damage was caused by the fact that a large part of New Orleans is below sea level.[18]

Again, it is very difficult to imagine that climate change is the primary cause of most of the world's flooding.

---

[18] https://www.carbonbrief.org/five-things-to-know-about-flooding-and-climate-change

CHAPTER 25

Forest Fires

Almost every year, it seems that we have huge forest fires in the West. In fact, one of the greatest was the 1910 fire that burned in Eastern Idaho and beyond. It destroyed the town of Wallace. So what causes forest fires?

Fires are caused by several things: carelessness, lightning, and arson. Some are caused by trees that fall on power lines. There is no evidence that forest fires are caused by a warmer climate, a climate that is one degree warmer. Forest fires are made worse when it is windy.

Most forest fires are made worse by poor maintenance. Forests need to be cleared of dead and dying trees that provide fuel for fires. Letting herds of animals work on forests also keeps the ladder fuel down so they do not set trees on fire. Trimming trees around the bottom to cut out ladder fuels is also a tremendous help in preventing forest fires.

Even clear-cuts can affect forest fires by providing large areas that cannot burn.

It is rare to see forest fires on private or state lands because they take better care of their forests. Forests burn because of lack of maintenance, not because of global warming.

Dr. Bob Zybach is an independent forest scientist based in Western Oregon. He has been the program manager for an educational Web site www.ORWW.org for twenty-one years as of January, 2018. His

PhD in environmental sciences is from Oregon State University (2003). His research interests include fire history, reforestation planning, forest management, Oregon Indian history, and forest history. This is what Dr. Zybach had to say about forest fires:

> I have researched the history of Pacific Northwest wildfires for more than 40 years and received a PhD from OSU (Oregon State University) regarding my 500-year history of catastrophic wildfires in western Oregon. Conditions are NOT "hotter and drier" than the 1930s, 1880s, etc., etc. Why politicians and academics keep making these easily disproven false claims and why the media keeps reporting them as facts can only be politics. In medieval times the high priests threatened people that they would burn in hell if they didn't leave money at the altar; today the threat is hell on earth if we don't pay carbon taxes in order to keep government bureaucrats employed. Same strategy, similar effectiveness, and neither threat seems to be based on actual facts or observations. Not sure if it's an actual hoax, but it certainly has turned into a lucrative racket. Bob Zybach
>
> Last month I (Bob Zybach) [October 31, 2017 Ed.] wrote a magazine article regarding this year's catastrophic wildfires in Western Oregon. The article mentioned that in the past, forest fires have killed thousands of people—the 1871 Peshtigo Fire in Wisconsin, for example, killed an estimated 1,500 to 2,500 people. Today, people not actively engaged in firefighting seldom die in forest fires, due in large part to modern transportation and communication systems.
>
> Within days of the article's publication, wildfires in Northern California killed more than

40 people, burned nearly a quarter of a million acres and destroyed more than 5,700 homes, businesses and other structures. This is the most people ever killed in West Coast wildfire history.

The California wildfires of 2017 are a true catastrophe, whether measured in acres, dollars, or human lives.

Western Oregon forest fires also burned more than half a million acres this year. The fires cost millions of dollars to fight, killed millions of wild animals, and subjected most of the state to foul-smelling, unsightly, unhealthy, choking smoke in August and September. But no one was killed during these events.

Compared to landslides, hurricanes, floods, earthquakes and volcanic eruptions, the loss of 40 lives might seem relatively minor. For families directly affected, though, it is a true catastrophe. When added to the destroyed homes, jobs, mementos, neighborhoods and communities, the loss must be nearly unbearable.

What can we in Western Oregon do to help rebuild Northern California's homes and communities, and to help reduce the increasing frequency, size, costs and destructiveness of these events? The best answers likely rest with the Department of Agriculture and the Department of the Interior. Lands managed by those agencies are where the large majority of fires are taking place, and those lands hold most of the dead trees that will likely fuel future fires if they are not removed.

Last month Interior Secretary Ryan Zinke and Agriculture Secretary Sonny Perdue joined forces in directing their agencies to "prevent and combat the spread of catastrophic wildfires through robust fuel reduction and pre-suppression techniques."

These statements were made before the California wildfires took place. Much of their focus was on the controlling and suppressing of wildfires, but "robust fuel reduction and pre-suppression techniques" imply long-term solutions via active management of our forests. That would signal a major shift from how many federal lands and wildfires have been managed during the last 30-plus years.

From 1951 until 1987, only one forest fire in Western Oregon exceeded 10,000 acres. Since 1987 there have been dozens of such fires, including 10 this year alone. Almost all of these fires have been on federal lands. Very few large fires in the past 70 years have burned on private or state lands—and those few were mostly affected by adjacent burning federal lands.

The principal difference is that private lands are actively managed for use and protection of their resources, while federal lands have increasingly become passively managed, allowing "natural processes" to take place with little or no human interference.

Active management in forested environments involves timely marketing and salvage of dead and dying trees; selective thinning of some areas; clearcutting, prescribed burning or reforestation in other areas; good road access; maintaining native wildlife populations and providing recreational opportunities. From World War II until the 1980s, most federal forests were actively managed to provide for national defense, post-war housing, wildfire control, public recreation and wildlife habitat. There were only a few large-scale wildfires during those years.

Passive management of federal forestlands largely began with the 1964 Wilderness Act, creation of the Environmental Protection Agency (EPA) in 1970, and the 1973 Endangered Species Act (ESA).

This process accelerated in following decades with government land-use designations, burgeoning ESA and EPA policies, and public litigation resulting in roadless areas, spotted owl habitat, and other large preserves in which active forest management was discouraged or even outlawed.

Recurring large-scale and catastrophic wildfires in these areas has become a predictable result, beginning in 1987 and growing worse since then.

Quickly salvaging trees killed in this year's fires would produce thousands of direct and indirect jobs, greatly improve the economics of rural Oregon and California, rebuild infrastructures harmed by the fires, reduce future wildfire risks and costs, and help provide economical, high-grade construction materials to restore ruined homes, businesses and communities. Safer, more beautiful forests for both people and wildlife would be another important result.

There are humanitarian reasons for shipping finished lumber and other materials to Northern California as quickly and cheaply as possible. There are sound economic and environmental reasons for doing so as well.

Bob Zybach of Cottage Grove, a forest scientist with a doctorate in the study of catastrophic wildfires, is program manager for the Oregon Websites and Watersheds Project (www.ORWW.org).

This was originally published as an editorial in the Eugene Register-Guard in October, 2017.[19]

(Used with permission.)

(For the story of the Paradise, California, fire, see the chapter on the "Paradise Fire in California," Appendix 6, also by Bob Zybach.)

---

[19] https://www.registerguard.com/rg/opinion/36070698-78/help-california-rebuild-by-managing-our-forests.html.csp

CHAPTER 26

Fracking And Modern Oil Production

Fracking, or hydraulic fracturing, is the process whereby oil wells are drilled quite deep through shale rock. Then the drilling goes horizontally. Finally, liquid, mostly water, is forced between the layers of rock under high pressure, and sand grains are left behind to keep the pores open. This allows them to bring oil and gas to the surface. Often, the pressure from the gas is enough to bring the liquid to the surface.

Now a film has been made, *Gasland*, which contends that fracking is polluting the groundwater. In the film, a faucet is turned on, and then gas is lit to prove that this is coming from the well. But upon inspection, it was discovered that natural gas was coming to the surface in these areas even before the wells were drilled. Those who make it their business to know, however, the regulators and scientists have demonstrated that most of the film is pretty misleading.

You can watch the film yourself. The movie is about an hour and a half.

First, here is the movie itself, *Gasland*. [20]

Fortunately, a woman decided to check it out. What she found out is a far different story, after talking to many professionals who knew the truth. She has produced this movie *Truthland* to help us understand the truth.

---

[20] https://tubitv.com/movies/310497/gasland

Here is some text and a link to an answer to *Gasland*. [21]

This film, *Truthland*, is contained in the above as a link. It is only about half an hour. The author spends a lot of time talking to the professionals. She concludes that *Gasland* is not only misleading but simply dishonest. [22]

The importance of fracking is that it enabled the United States to go from a net user of fossil fuels to a net shipper. We now have more oil and gas than any other nation on Earth. It continues to be important to point out that coal, oil, natural gas, and nuclear power have enabled America to become the greatest nation on earth. If we abandon fossil fuels to meet a very risky future with solar and wind, we may expect a future that is far different than we now enjoy. Fossil fuels are what made America great, but wind and solar may be the means of putting us back in the Stone Age and cause us to live in darkness.

Here is a delightful short film about fracking made by The Heartland Institute, with Isaac Orr: What is fracking?[23]

Oil wells with fracking are very deep, some over a mile. The wells are sealed off with several layers of casing to prevent any oil or gas getting away.

Man is not the cause of atmospheric warming that is going to cause the end of the world as we know it. There are many benefits to fossil fuels, and we need to recognize this before setting out on a program to convert to only wind and solar, without knowing either the cost or the benefits.

---

[21] https://www.hotair.com/archives/2012/07/01/video-truthland-is-the-answer-to-gasland/

[22] https://www.youtube.com/watch?v=iTJaaeiuzSU

[23] https://www.youtube.com/watch?v=UWaFo8tfhb8

CHAPTER 27

Gasohol

Have you ever pulled up to a gas pump, gotten out of the car, then noticed there was a note that said the pump was out of order? So you had to move to another pump. Or you lived during the time of gasoline rationing during the 1970s. At that time, gas-guzzling motor homes almost disappeared from the highway. Gasoline went to $4.00 a gallon.

During that time, we were driving from Idaho to Pensacola, Florida. We pulled into a gas station to get gas somewhere in the South and were told they were out of gas. We asked to use the restrooms, and while we were there, they served several customers. They were not out of gas; they were only saving it for the locals.

We finally got gas somewhere and went on our way. As we approached Houston, Texas, we just knew we would not have any trouble as that is the gasoline capital of the world. But as we got closer and closer, we saw very long lines of cars at gas stations.

So we found a campground and parked our trailer. The next morning, I got up while it was still dark. Quite close, I found a small gas station that was willing to sell us gasoline. He said he would sell as long as he had gasoline. So after that, we never let the tank get lower than half full.

Gasohol is a gasoline that has ethyl alcohol (or ethanol) added, usually no more than 10 percent. Alcohol contains less carbon so less carbon dioxide when it burns, so burning gasohol was supposed to mean less carbon dioxide in the air. But there is less power in gasohol, and it is harder on internal combustion engines.

Gasohol is made from corn, and many farmers started raising lots of corn, especially in the Midwest states. But making gasohol meant less corn for eating. This created problems in some areas of the world, and there were near riots over food.

Growing corn and other grains for ethanol productions requires a lot of farmland, taking up land that could be used to feed the world's hungry. Also, the raising of corn takes a lot of fertilizer and herbicide which can lead to nutrient and sediment pollution.

Carbon dioxide is one of the by-products of alcohol production; therefore, the question has to be asked if the carbon dioxide we save as a motor fuel is more than is created in the production of the alcohol in the first place. Also, there is the carbon dioxide produced from the many tractors that plow up the land and harvest the corn. According to some experts, the production of corn-based ethanol as an alternative fuel may end up requiring more energy than the fuel can generate, especially when counting the high energy costs of synthetic fertilizer production, (And consider that most fertilizer is made from natural gas which would be eliminated under the Green New Deal.)

Right now, we have lots of oil, so using gasohol probably makes no sense, except that a number of people in some states receive a subsidy for the production, and now it would be hard to stop.

The production of gasohol is just another example of national policies that do not change with changing times. No matter, man is not the cause of global warming, and it is very probable that using gasohol does not do anything to lower the world temperatures.

CHAPTER 28

Georgetown, Texas, And 100 Percent Renewable Energy

Georgetown, Texas, is very proud of being one of the first cities to go 100 percent renewable energy.

A 150-megawatt solar power agreement finalized in 2015, in addition to a 144-megawatt wind power agreement in 2014, will make the city of Georgetown one of the largest municipally owned utilities in the United States to supply its customers with 100 percent solar and wind energy. The long-term agreements also allow Georgetown to provide competitive electric rates and hedge against price volatility for energy produced by fossil fuels.[24]

The 1,250-acre Buckhorn Solar Plant located fifteen miles north of Fort Stockton contains 1.7 million solar panels. The panels are mounted on a single-axis tracking system that rotate over the course of each day to maintain a ninety-degree angle to the sun in order to maximize output. The plant is in Pecos County, which has the second-highest radiance factor in the state. Radiance is a rating for available sunlight.[25]

The 150 megawatts power was to go into effect in 2017.

---

[24] https://www.gus.georgetown.org/renewable-energy/
[25] https://www.georgetown.org/2018/06/29/georgetowns-energy-100-percent-renewable-with-solar-plant/

There is still no discussion as to what they will do when the sun is not shining or the wind is not blowing.

While Al Gore was pretty excited about the situation in Georgetown, Texas, all was not so rosy. Political leaders in this college town in central Texas won wide praise from former Vice President Al Gore and the larger Green Movement when they decided to go 100 percent renewable seven years ago. Now, however, they are on the defensive over electricity costs that have the residents paying more than $1,000 per household in higher electricity charges over the last four years. This was actually $1,219 per household in higher energy costs for the 71,000 residents of Georgetown, Texas, all thanks to the decision of the mayor and the council. He had a bold plan to shift the city's municipal utility to 100 percent renewable power in 2012 when he was on the city council. (This was probably $1,000 more per year Ed.)

By 2017, the mayor's green gamble was undercut by the cheap natural gas prices brought about by the revolution in high-tech fracking. Power that year cost the city's budget $9.5 million more than expected, rising to $10.5 million in 2018, according to the *Williamson County Sun*.

That a city in Texas claimed to be 100 percent renewable generated significant notoriety. But as University of Houston energy expert Charles McConnell noted, "It's not kind of misleading. It's very misleading, and it is for political gain."

Study as you will, you find no discussion of what they will do when the sun is not shining or the wind is not blowing.[26]

And of course, man is not the cause of global warming, so all this is a wasted effort.

---

[26] https://www.wattsupwiththat.com/2019/01/29/epic-fail-of-renewables-causes-texas-town-to-have-1200-per-month-higher-power-bills/,
https://www.foxnews.com/opinion/texas-towns-environmental-narcissism-makes-al-gore-happy-while-sticking-its-citizens-with-the-bill

CHAPTER 29

Green New Deal Is Socialism

The one thing that almost all socialist countries have in common is the required use of force. During the last century, millions lost their lives under socialism, be it Nazism, Russian Communism, China, or others. People want freedom, and if they can't get it, they leave, or they are shot.

Take a recent example, Venezuela. It is now a socialist country and the people are leaving in large numbers; it was leave, or go hungry. The country was even trying to keep out any aid that was to come to them. Maybe by the time this is published, many of them will be dead.

Few of our young people today know anything about socialism, or about capitalism for that matter. They are taught little about economics in school, and even our professors in college seem to have little understanding of economic systems.

Therefore, when they hear a Bernie Sanders or an Alexandria Ocasio-Cortez talk about changes that amount to socialism, they have no frame of reference with which to reject them.

Now we have the Green New Deal. They feel the first New Deal was a great success, when in fact it got us started on the road to socialism.

It is surprising the number of congressmen who agree with Ms. Ocasio-Cortez. They have been around long enough that they ought to know better, but it seems that they do not.

Let us look at this Green New Deal. One writer (Rich Lowry) said that there's nothing the "Green New Deal" can't do.

He said further, "This isn't like a European country adopting an ambitious goal for renewables (Denmark wants to be at 50 percent by 2030); it is a country (the USA) with more recoverable oil reserves than Saudi Arabia and Russia spurning a stupendous source of national wealth to take a flyer on a lunatic experiment."[27]

So what is their plan? What are they hoping the American people will go for?

1. 100 percent renewable sources of all national power in just ten years.
2. Since renewables (wind, solar, hydroelectric) provide only 15 percent of our power, it would require the shuttering of 85 percent of American power plants.
3. Rebuild or replace every home in America (136 million homes). They have no idea how much lumber this would take, as well as insulation and other resources, and what affect this would have on the environment.
4. Eliminate all emissions from industry and farming.
5. Eliminate all emissions from transportation, which would mean the use of all electric cars and trucks, and possibly hydrogen for aircraft (if that is even possible). Electric cars have a range of only about three hundred miles and far fewer in winter and summer. Big trucks would have to take up half their space with batteries. Where will we get so many batteries?
6. A living wage for all, even if they do not want to work.
7. Mandatory union membership, even if they do not want to join a union. They would have no choice.
8. Universal health care.

---

[27] Rich Lowry, The "Green New Deal" Isn't Just Ambitious—It's Insane, https://www.wattsupwiththat.com/2019/02/13/laborers-international-union-of-north-america-savages-aocs-green-new-deal/

9.   Use of high-speed rail in the place of aircraft. What about Hawaii and other faraway places like Alaska?
10. Ban nuclear, the most concentrated energy on Earth.
11. End all traditional uses of energy in just ten years. No coal, oil, and natural gas. These are very concentrated hydrocarbons, and wind and solar will never take their place, no matter how much we wish they could.

So this is their program. It is designed to restrict the temperature rise to just 2°C. But what they seem to be missing is just two things: First, no one has ever been able to prove that the things above are actually causing climate change or global warming, and second, fossil fuels have made America the greatest nation on Earth in just a little under two hundred years.

Do we want to go back to sailing ships and living like our ancestors lived? Most Americans are happy with the way we live and have no idea what these proposals are going to mean to them personally. Even if you could somehow prove that man is the cause of global warming (and you can't), we still have no way to know if the Earth would be destroyed if the temperature were to rise above the limit of two degrees. Historically, the Earth has been far warmer and far cooler at times in the past. What would happen to all the climate conferences held throughout the world if there was no air traffic? It is more like shooting themselves in the foot.

No matter how you figure it, the Green New Deal is not the solution to anything, and man is not the cause of worldwide global warming.

CHAPTER 30

Greenland And Glaciers

Much has been said in recent years of what would happen if Greenland should melt and all that water flow into the sea. The problem with this thinking is twofold. First, Greenland is not a flat plain covered with ice, but is mountains and valleys, so not all the ice would simply slide off into the ocean.

The topographic map of the Greenland ice cap shows a broad dome, leading to the common assumption that the ice cap is underlain by a dome-shaped continent. A map of the base of the ice shows this is not true. Instead, a kilometer-deep basin extends below sea level under the Greenland interior, the sub-ice terrain being a bowl formed by a ring of mountains with few openings to the sea. This results from the mass of the ice cap being heavy enough to cause isostatic (where one area rises, another sinks) sinking of the land.

Importantly, therefore, the ice cannot simply slide into the sea as is often alleged. Instead, ice near the base of the ice cap flows upward to join glaciers flowing through gaps in the mountain rim. According to the IPCC *Fourth Assessment Report*, melting of the whole ice sheet would contribute nearly seven meters to sea level rise.[28] Yet if the whole ice sheet could suddenly melt, much of the water would be retained in a huge lake bounded by the mountain rim. In any case,

---

[28] Bergmann et al., 2012

the distribution of annual mean temperatures on Greenland is such that melting is possible only around the periphery.[29]

Greenland is one of the coldest places in the world.

Second, Greenland has been warmer in the past, and it has not melted.

Przybylak (2000) published a comprehensive meteorological analysis that provides strong support for Joughin's (reference same page) conclusion, stating, "The level of temperature in Greenland in the last ten to twenty years is similar to that observed in the nineteenth century," and citing corroborating evidence for an earlier warm Arctic in the 1930s and 1950s. Przybylak concluded the meteorological record "shows that the observed variations in air temperature in the real Arctic are in many aspects not consistent with the projected climate changes computed by climate models for the enhanced greenhouse effect, because 'the temperature predictions produced by numerical climate models significantly differ from those actually observed.'" These conclusions are supported by Greenland temperature records dating back to 1880.

The studies discussed so far fail to take adequate account of the Holocene context within which modern glacial change must be considered. The record indicates warmer temperatures were the norm in Greenland in the earlier part of the past four thousand years, including century-long intervals nearly 1°C warmer than the recent decade of 2001–2010 (Ibid. p. 642).

The point of this reference is that it has been warmer in the past, and Greenland has not simply melted. Like most of the world, Greenland has been warmer and colder during the past several thousand years.

So who are you going to believe, the countless man-made models or the climate temperature record? The temperature record has a tendency to make the models false.

The studies reported above make clear that any recent upswing in glacial outflow activity on Greenland has no relationship with anthropogenic global warming, as late-twentieth-century tempera-

---

[29] *Climate Change Reconsidered II: Physical Science*, pp. 641–642

tures did not rise either as fast or as high as they did during the great natural warming of the1920s–1930s (Ibid. p. 645).

As noted elsewhere, Greenland gets a lot of snow. It has snowed almost three hundred feet since the Second World War. This was proved simply by the fact that a P-38 airplane that landed on the ice during the war was covered by almost three hundred feet of snow when it was found in recent times. Any snow that comes to Greenland means water taken from the sea, so this would tend to keep the sea level lower.

It has to be evident that any possible warming in this century was not a part of global warming that may have been caused by the activity of man. Whatever warming has been noted was most probably caused by natural processes and not by man.

CHAPTER 31

History of Coal

The history of coal mining goes back thousands of years. It became important in the Industrial Revolution of the nineteenth and twentieth centuries, when it was primarily used to power steam engines, heat buildings, and generate electricity. Coal mining continues as an important economic activity today.

Compared to wood fuels, coal yields a much higher amount of energy per mass and can often be obtained in areas where wood is not readily available.

Historically, coal was used to heat homes, but now it is used primarily in industry, especially in the production of electricity and in the production of steel. (When we first moved to Idaho, our home was heated with coal. When we remodeled, we switched to natural gas.)

Coal remains a key industrial fuel because of it being found all over the world and because of the low cost. It is still one of the cheapest fuels for generating electricity. Most of the early coal mines were in Britain.

Today, coal has been replaced by oil and natural gas. Natural gas is used especially for electrical generation, especially because natural gas, or methane (CH_4), has a lot more hydrogen for each atom of carbon.

Coal remains a vital form of energy for the rest of the world. China and India are building hundreds of coal-fired power plants, and it has become important for Africa. At a time when third-world countries are just beginning to come out of the Stone Age with the use of coal and oil, environmentalists would take these vital energy sources away from them. With electrical energy, people live longer and have lives that are more fulfilling.

Coal was mined in China as far back as 3,450 BC.

America became the greatest industrial nation on the Earth primarily because of two things: (1) our constitution with our free enterprise system of law and (2) the use of our natural resources, coal, oil, and natural gas.

If we were to eliminate the use of coal, our nation would go back to living as we did before these things were discovered.

Our use of coal is not causing or going to cause global warming.

CHAPTER 32

History of Hydroelectric Power

Water was probably used to grind flour long before electricity was discovered. This goes back to about the second century B.C. They were what was called "vertical-axis mills" and were built and used throughout the ancient Greek and Roman empires.

A water turbine converts water power to electricity and was the first renewable source of electricity. Dams have been built all over the country but mainly in the West. Now there is talk of removing the dams to make it better for fish, but it is a foolish measure because dams provide not only electricity but flood control and water for irrigation.

In Europe, horizontal-axis water mills were later invented and were also used to grind grain.

The invention of the steam engine aided in grinding of grain and in the locomotive power of rail equipment.

Many inventors developed various water turbines, but it was not until the early twentieth century that Victor Kaplan invented the Kaplan turbine which provided for the modern usage of low-head water turbine generators.

Hydroelectric dams at this point in time (2019) are pretty limited as most all the good locations for dams have been taken. This includes states with dramatic changes in elevation in Idaho, Oregon, Washington, and South Dakota. This also includes Colorado and most other states where even a few dams are located. Tennessee has one of the first dams. Flat states like Florida, Ohio, and Virginia, do not produce much electricity.

In the early years, hydropower produced most of the electricity, but now it is only about 10 percent. However, hydroelectricity is still the number one source of renewable energy and is an excellent reason to keep them in operation. It makes no sense to spend millions building dams and then spending more millions tearing them down.

Today, most new projects use smaller streams and lakes to provide for individual communities.

Hydroelectricity provides about 20 percent of the world's electricity, including China, Canada, Brazil, and the United States. There are many great locations outside the United States that could still be used to build dams, if anyone wanted to use them.

Hydroelectricity still makes the most sense. Sure, some dams are a little unsightly, but think of all the opportunities of using the lakes they provide. These are still the best means of generating electricity. And they are great opportunities for recreation, boating and swimming.

Hydroelectric dams do not cause any aspect of global warming, and there are other great programs for dealing with the needs of fish. [30]

---

[30] http://www.turbinegenerator.org/hydro/history-hydroelectric-power/

CHAPTER 33

History of Nuclear Power

Nuclear power was used to make bombs for defense long before it was used to generate electricity. Nuclear power was first used for electricity generation in the 1950s. The world's first nuclear power plant for commercial electricity generation was at Calder Hall in Sellafield, Great Britain, and was completed in 1956. It produced steam for electricity generation as well as plutonium which was used for defense purposes.

From 1960 to 1970, nuclear power grew from about 1 Gw to over 100 Gw. Nuclear power grew in a move away from the use of oil and coal because of the oil crisis of the 1970s.

Criticism of nuclear power became greater after March 28, 1979, when the Three Mile Island nuclear power plant, near Harrisburg, Pennsylvania, in the United States suffered a series of technical errors (in reading the gauges) which resulted in a partial meltdown. Although one reactor was destroyed, no radioactive material leaked out, and no one was injured. Even so, this accident had a major effect on nuclear development and the debate over nuclear policy.

The problem at Three Mile Island was not serious, and America has never had a person killed in a nuclear accident. More people fall off roofs than are hurt by nuclear reactors.

However, people began to become afraid of nuclear power after the nuclear accident in Russia as well as Three Mile Island. The Russian nuclear accident happened on April 26, 1986, at Chernobyl in Pripyat in the Ukraine, part of Russia at the time. There appears to have been thirty deaths immediately, but many more were contaminated, and it is hard to know just how many eventually died from the accident. This was considered the worst nuclear accident in history.

One reason for the meltdown was that the Chernobyl reactor did not have a leak-proof containment structure surrounding the reactor, something that all existing power plants have today.

Another nuclear accident was the Fukushima Daiichi reactor meltdown. This was caused by an earthquake and a fifteen-meter 49.2 feet tsunami on March 11, 2011. This damaged the reactor so that it was unable to cool itself. A large area was evacuated because of radiation concerns. It was pretty dramatic watching the sea wash over land.

After these accidents, there was considerable pressure around the world to eliminate nuclear power. However, there have been many safety improvements on nuclear power plants so that they are much safer at the present time. Nuclear power plants have been used for ages on U.S. submarines and aircraft carriers without serious accidents.

Some countries (Italy, Germany) have closed down or are in the process of closing down their nuclear power plants, but it will be interesting to see what they use to take their place.

Public attitudes concerning nuclear power are changing so that there are a few new power plants in the process of being built. The newest nuclear power plant built in the United States is at Watts Bar, Tennessee, June 2016. Nuclear power is probably the most efficient power in the world and has no pollution.

If we happened to be concerned about global warming, then building more dams and more nuclear power plants would seem to be the way to go. It is obvious that using nuclear power plants has not caused more global warming.

CHAPTER 34

History of Oil

Discovery of oil in America

Oil is the energy that has made the civilized world. Before the discovery of oil, the world was still burning wood and some coal. But in early America, oil was collected from seeps, such as the La Brea Tar Pits in Los Angeles. (The La Brea Tar Pits has many ancient fossils, some extinct.)

Then in 1859, a man named Edwin Drake drilled the first well and hit oil at 69.5 feet. This was at Cherrytree Township, in Venango County, Pennsylvania. In just a few years, others copied his methods, and there was lots of oil. This was far better than the traditional oil for lamps taken from whales, and it also helped protect the whales from extinction.

Oil was actually discovered before the Drake well, by people who were drilling for salt brine, but hit a layer of oil. ("Curse of Texas" as they said in the movie *Texas Across the River*).

Drake (*right*) in front of the well
(This well house burned down but was rebuilt. In 1945, a
replica (on right) including pump was built on the site.)

Once oil was discovered, many new uses were found for it. One of the first was kerosene, which was used for "coal-oil lamps." Later, the oil was broken down into all sorts of chemicals.

Some of the more common things that come from oil are gasoline and diesel, kerosene (now used for airplanes), and tars for asphalt and surfacing of roads, roofs, and floors. Plastics also come from oil, from which a multitude of things are made. (How would you like to brush your teeth with a wooden toothbrush?)

Some other things you might not think about are cosmetics, lubricants, tires, medicine, cleaning products, paints, and fabrics. If we stop the use of all fossil fuels, are we willing to do without all these things? Probably not!

The discovery of oil changed America forever. Now no longer dependent on wood, people could use oil to power all sorts of machines. Coal took the place of wood in the operation of locomotives, and later diesel oil with diesel/electric power. Coal was also important in the generation of electricity. Then natural gas took the place of coal in many electrical generation plants.

Natural gas is the simplest carbon compound CH_4 and is usually found along with the oil. It is the gas that occurs with oil that sometimes forces the oil to the surface.

The importance of the Drake well is that it encouraged a lot of investments and oil became important as an industry.

The Competitive Enterprise Institute has done a lot of work on energy, and a number of their leaders have written important books on the subject.

The executive summary of one of their most recent papers, "Energy White Paper," January 23, 2019, had this to say (Myron Ebell, Marlo Lewis Jr., Chris Horner, and Ben Lieberman):

> Energy is the lifeblood of the economy. Thanks to affordable energy, the average person today lives longer and healthier, travels farther and faster in greater comfort and safety, and has greater access to information than did the privileged elites of former times. Carbon fuels—coal, oil, and natural gas—provide 80 percent of U.S. energy and 87 percent of global energy. They are the world's dominant energy sources because, in most markets, they beat the alternatives in both cost and performance.

The EPA and the Clean Power Plan have worked for decades against that which is best for Americans. The Competitive Enterprise Institute (CEI) has this to say about the plan:

> The Clean Power Plan has legal flaws beyond the Court's errors in Massachusetts v. EPA, and the agency is currently in the process of repealing it. Nonetheless, as long as Congress treats Massachusetts v. EPA as settled law, future executives will be tempted to usurp legislative power.

If you are interested in the preservation of energy policy, you need to read the "White Paper" Footnote (link) 31 below. The CEI had more to say:

> The wealth creation and technological progress made possible by affordable carbon based energy

95

make societies more resilient, as they protect people from extreme weather, power health-improving innovation, and increase life expectancy. Since the 1920s, global deaths and death rates from extreme weather have decreased by 93 percent and 98 percent, respectively.

Not only does oil (and coal and nuclear energy) make life better, there is real benefit to the way we live our lives. Read what CEI had to say:

The war on affordable energy also raises serious humanitarian concerns. Energy costs already impose real burdens on low-income households, including reduced expenditures for food, medicine, and education and late credit card payments.

"Consensus" climatology implies that the Paris climate treaty's objective of limiting average global temperatures to 2°C above preindustrial levels cannot be accomplished without massive cuts in developing countries' current consumption of carbon-based fuels. Putting the developing world on an energy diet is bound to be a cure worse than the supposed disease.

Here is more on the subject:

Today, critics claim that unchecked carbon energy use will cause catastrophic climate change. However, the climate models producing scary impact assessments project about twice as much global warming as has actually occurred. More important, the climate change mitigation policies those critics advocate pose serious risks to American prosperity, competitiveness, and living standards.

The Green New Deal calls for a higher tax on the rich which has never produced any good. In fact, it is the so-called rich who are providing most of the jobs that exist in the world today. No more rich, no more jobs.

With some plans, they expect to tax gasoline then give that back to the people in general. Most people who have to travel will just spend the extra income on transportation, and it becomes a circular solution. So how is this any help to the challenges of global warming?

Read the "White Paper," and you will be glad you did.[31]

Just a few years ago, the United States had to buy most of its oil abroad, especially from countries that don't like us very well. Then in about 1950, we discovered how to drill wells and extract oil by a process called fracking. They often drill in an ocean that is two miles deep, then drill through shale for another mile, then drill horizontally another mile. When they are through, using very high pressure, they force liquid between the layers of shale, then leave grains of sand behind to keep the spaces open, and let the oil out. As a result, the United States is now a net producer of oil. We produce more oil than any other country and even ship oil abroad.

Who knows what the next big advance in oil production will be, but it is going to be interesting to see. Oil is vital, and when we can produce it cheaply, then green energy has no advantage; in fact, green energy has many disadvantages.

Oil exploration in the Mideast

After the discovery of oil in Pennsylvania, geologists began to look for oil all over the world. Ultimately, the Middle Eastern countries offered the most promise, so work began there to find oil. Most of the work was done where there were already oil seeps, so wells were started near those seeps.

---

[31] http://www.masterresource.org/competitive-enterprise-institute/cei-energy-environmental-policy-2019/

In March 1901, William Knox D'Arcy sent his agent Alfred L. Marriott with Cotte and Kitabchi Khan to Iran to negotiate a concession from the Persian (Iran) government.

D'Arcy worked out an acceptable agreement and began drilling. Ultimately, he developed some very profitable wells and the world has not been the same since.

The discovery and development of each one of these oil fields is a very exciting story in itself. Until very recently, the United States was totally dependent on these foreign oil fields for the oil we use, but these countries have not had good refineries, so they have had to purchase gasoline from the United States.

Major oil fields have been developed in Iraq, Iran, Saudi Arabia, as well as other parts of the world, such as Venezuela, not to forget such places as Bakersfield, and Long Beach, California. Unfortunately, most of these great foreign oil fields have been taken over by the host countries.

Judging by our history and the levels of carbon dioxide, it is easy to conclude that using oil has not been the chief cause of global warming.

CHAPTER 35

Home Solar Panels

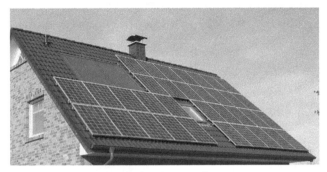

Home with a large solar array

Home solar panels seem like the ideal. They produce electricity, and the homeowner is paid a certain price. In some places, the owner is even given a grant to encourage the installation of the panels.

Home solar panels make a lot of sense when the home is far enough south to catch a lot of the sun or where there are not a lot of cloudy days. They make no sense farther north.

Of course, home solar panels are a real problem when the time comes to replace the roofing, as they must first be removed before the roof can be replaced.

Some folk have used solar panels to heat water for a preheater for the water heater. This can be a real benefit. Or they can be used to heat a swimming pool.

Home solar panels make no sense when the homeowner has to pay for the panels himself. The panels will be worn out before he earns enough from the electricity generated to pay for the panels.

Home solar panels also attract a lot of dust which has to be cleaned off periodically. It is the same with leaves that cover them in the fall.

Home solar panels make the most sense in locations where the home is far from the electrical grid. We used them to charge a battery when we lived in Mexico, far from the electrical grid. We charged a battery which we used for light in the house. But to wash clothing, we had to use a gas-fired generator. It would take a lot of panels and batteries to have enough power for a washing machine, and it is just not worth the cost of the installation.

Using home solar panels does not help prevent global warming.

CHAPTER 36

Increasing Carbon Dioxide Rate (Is It True?)

Carbon dioxide is the gas of life. All plants on Earth get their substance from the minute quantity of carbon dioxide in the atmosphere. Animals eat the plants so you could easily say that all life gets its substance from the carbon dioxide in the air.

Carbon dioxide is gradually increasing. Today (2019), it is about 0.041 percent. That is really not much. To keep in perspective just how much carbon dioxide there really is one needs to look at what makes up our atmosphere. Here is the list:

Nitrogen   78%
Oxygen   21%
Argon   1.0%

The rest are not even worth mentioning, but here they are, just a trace of these gases: Water is probably more than the rest but also CO_2, neon, helium, methane, krypton, hydrogen, nitrous oxide, xenon, ozone, carbon monoxide, and ammonia. Some of these are in such small quantities that it is almost impossible to detect them. Carbon dioxide is just one of them.

The station measuring carbon dioxide is located on Mauna Loa, in Hawaii, but since the Big Island (Hawaii) is a volcano and has at least one active volcano (Kilauea), it would not seem the best place for this instrument, as volcanoes give off a lot of carbon dioxide.

It is important to note, however, that carbon dioxide always increases in winter and goes down in summer, when plants are growing. Carbon dioxide has been increasing at about the same rate for the past one hundred years, even as temperature has gone up and down, and for the past decade, the temperature seems to have leveled off (temperature pause).

There has been quite a bit of discussion putting forth the idea that the rate of increase of carbon dioxide is increasing, but there is no evidence for this. Even if it were, it would not be a serious problem. Plants do much better with higher concentrations of carbon dioxide, and people live on submarines and spaceships with much higher concentrations of carbon dioxide. There is no evidence that the rate of carbon dioxide is increasing, and it would be a good thing if it did.

There is no evidence that a one-degree increase in temperature is having a radical effect on the level of global carbon dioxide and global warming.

CHAPTER 37

Katrina and Hurricanes (2005–2017)

Katrina (of 2005) had a number of serious problems. First, New Orleans had no plan as to what to do if they were actually flooded. Much of the town of New Orleans is below sea level. Thus, if the pumps failed, the town would gradually be filled with water.

Second, there were over a hundred school buses sitting in the water, which could have been used to pilot people to safety had they been organized to do so, and had there been drivers with permission to drive them to safety.

The third thing is that many of the residents of the town had been conditioned to believe the government would take care of them, and as the water began to rise, they just assumed that someone would come to rescue them. But no one came except a few people with boats or some who "borrowed" boats.

Then, the Astrodome, which was on higher ground and held a lot of people, had no way to handle the sewage. Pretty soon, the toilets quit working.

There was also the town across the bridge which refused to help them simply because they had all they could do just to help their own people. A total of 1,833 people died in Hurricane Katrina.

Here is a list of the major hurricanes for the 2005 season:

Dennis	July 10	Category 4
Emily	July 14	Category 5
Katrina	August 23	Category 3
Rita	September 26	Category 3
Wilma	October 24	Category 3

Here is a good place to talk about what the categories mean. This is the Saffir-Simpson scale, named after the two men who developed it. Here is the scale:

Category 1	74–95 mph	3–5 ft storm surge
Category 2	96–110	6–8 ft storm surge
Category 3	111–130	9–12 ft storm surge
Category 4	131–155	13–18 ft storm surge
Category 5	155 or more	more than 18 ft storm surge

The 2005 hurricane season was so severe that some climatologists predicted even worse hurricane seasons for the future. But the next serious hurricane was Hurricane Harvey, August 25, 2017, which dumped multiple feet of rain on Houston, Texas, then Maria, category 4 in September 2017. For twelve years, between 2005 and 2017, there were no hurricanes category 3 or above. All that time, carbon dioxide levels continued to rise, so, obviously, hurricanes are caused by something other than carbon dioxide.

The 2005 Atlantic hurricane season was the most active Atlantic hurricane season in recorded history, shattering numerous records. In response, articles appeared in relation to climate change and hurricanes. *Time Magazine* published an article titled, "Is Global Warming Fueling Katrina?" Shortly after the hurricane, former *The Boston Globe* reporter Ross Gelbspan wrote an op-ed piece for *The Boston Globe* titled, "Katrina's Real Name," declaring that the hurricane's "real name is global warming" (Wikipedia).

Hurricanes do a lot more damage today mainly because there is a lot more development in their path. Back in the days of adobe buildings, they did not do much except to flood the home. Now, they blow a lot of roofs away and soak the buildings.

Still, as mentioned, we have had only a few hurricanes while the carbon dioxide level was rising, so they must not depend on carbon dioxide. And as far as greenhouse gases, it is important to note that of all the greenhouse gases, water vapor is the most abundant by far and the amount varies greatly from day to day and from place to place.

It is evident that man is not causing hurricanes by using fossil fuels.

CHAPTER 38

Ocean Acidification: What Does It Mean?

Not long ago, I saw a picture of a professor with a sea animal and a partially dissolved shell. Sea animals have shells that are made of calcium carbonate (limestone), and if placed in an acid solution, they will dissolve.

But I wondered why I continued to see seashells on the beach if the sea was too acidic. Was this the case everywhere, or was this just an isolated example?

I concluded that the professor's example did not apply to all sea animals.

First off to talk about sea acidification is a misnomer. The sea is not becoming more acidic, but it is becoming less basic. The pH scale runs from 0 to 14 with 7 being neutral (pure water). Anything greater than 7 is basic, or alkaline, and anything lower than 7 is acidic.

So what is the pH of the ocean? Traditionally, the ocean has been at about pH 8.2. Today, it is about 8.1, slightly less basic.

Some scientists have grown seashells in water that is slightly acidic and found that they do not grow properly, leaving them with soft shells, etc.

Seashells are made from calcium and carbon dioxide to form calcium carbonate or calcite (limestone).

The calcium comes from sea animals that have died, and limestone is found in rocks.

The oceans would have to become a lot more acidic before there would be a problem.

This obviously is not caused by global warming or man's activity.

CHAPTER 39

Ocean Warming

Some scientists believe that the oceans are warming faster today than previously thought. This would seem to have dire implications for climate change because all the excess heat is stored in the oceans, which hold fifty times more heat than the atmosphere.

Because the oceans hold much more heat than the atmosphere, whenever our atmosphere warms up, the oceans gain most of that heat without very much warming. Without the oceans, the land would warm up much faster. Of course, the amount of water in the atmosphere, the major greenhouse gas, keeps the temperature fairly even and keeps it from being either too hot or too cold. The air in the Tropics holds the most water vapor, so there the temperature has a tendency to hold to a fairly consistent temperature, both night and day, winter and summer.

Because the oceans hold so much heat, they save us from great warming right now. The Earth continues to warm and has done so since the end of the last Ice Age, about ten thousand years ago. Most of this change in warming has probably come from the changes in the energy of our sun.

At one time the temperature of the oceans was taken by passing ships, who measured the temperature. Since the early 2000s, scientists have measured ocean temperature using a system of floats called Argo, named after a ship in Greek mythology.

One thing to keep in mind is that as water warms, it expands, and this raises the level of the ocean. So as the ocean is gradually warming, the sea level is gradually rising. As far as local sea level, sometimes, it is not the sea rising but the land subsiding or getting lower.

It is important to note that the sea is warmed mainly by the sun, and because the oceans are very deep, it takes thousands of years to warm to the bottom. Man has only been burning fossil fuels for about one hundred years, so this could not be the cause of global warming, which is probably caused by the sun or other forces of nature.

CHAPTER 40

Oil Discovered in Alaska

Oil was first discovered in Alaska in 1957. The discovery was on the Kenai Peninsula and produced nine hundred barrels a day.

Oil was later discovered on the North Slope (above the Arctic Circle, next to the Arctic Ocean) on March 12, 1968. This was the Prudhoe Bay Oil Field and included about twenty-five billion barrels of oil (a barrel is forty-two gallons). Prudhoe Bay is four hundred miles north of Fairbanks, Alaska. The field produced 1.5 million barrels a day.

The presence of oil on the North Slope of Alaska was suspected for more than a century. In 1968 the Atlantic Richfield Company and Humble Oil (now Exxon) confirmed the presence of a vast oil field at Prudhoe Bay. Within a year, plans were under way for a pipeline.

In 1970, environmental groups and others filed suits to prevent pipeline construction. Three and a half years of legal proceedings followed, during which the proposal to build the pipeline was considered by the federal and state governments, including the US Congress. No construction was permitted during this time.

Presidential approval of pipeline legislation provided the go-ahead to begin construction on November 16, 1973. The 360-mile distance from the Yukon River to Prudhoe Bay required a road to be built for transportation of equipment and materials. It was constructed in 1974. At the same time, work was begun on pump stations, the pipeline work pad, and the Valdez Marine Terminal.

Pipeline employment reached its peak at 21,600 in August of 1975. By May of 1977, all eight hundred miles had been installed and tested. Oil entered the pipeline at Pump Station 1, at Prudhoe Bay, on June 20, 1977, and reached Valdez on July 28. On August 1, 1977, the tanker "ARCO Juneau" sailed out of Valdez with the first load of North Slope crude oil.

In some places, the pipeline was insulated and suspended, and in other areas, it was buried, depending on the soil conditions.

The Valdez Marine Terminal

Oil from the pipeline is first stored and then loaded aboard tankers at the terminal at Valdez. Located across the bay from the city, this one-thousand-acre site is built at the northernmost year-round ice-free port in the United States.

There are eighteen crude-oil storage tanks at the Valdez Marine Terminal. The tanks are 250 feet in diameter, 62 feet high, and can hold 510,000 barrels each, for a total capacity of 9.18 million barrels. Each is surrounded by a concrete dike capable of holding 110 percent of the oil in the tank, in the event of a spill.

The nerve center for the entire eight-hundred-mile-long pipeline is the Operations Control Center located at the Valdez Marine Terminal. The controllers at the center can start or stop the entire pipeline or initiate or terminate functions at any part of the line. There are also other facilities at the Valdez Terminal to support all the work being done.

Tankers arrive almost daily at Valdez to carry crude oil from the pipeline terminal to refineries to the south.

The Exxon-Valdez oil spill

The Exxon-Valdez Oil Spill took place in Prince William Sound on March 24, 1989. The ship struck a reef and the oil spilled. The tanker was carrying 53,094,510 gallons and spilled about 10.8 million gallons of crude oil. This was one of the most devastating human-caused disasters in the American history. It was second only to the 2010 Deepwater Horizon oil spill in the Gulf.

The spill killed 100,000 to 250,000 seabirds, at least 2,800 sea otters, approximately 12 river otters, 300 harbor seals, 247 bald eagles, 22 orcas (whales), and an unknown number of salmon and herring.

Cleanup was begun immediately, but this was difficult because there were no roads to the area and it was hard to get to. Some oil was caught, and chemicals were used to dissolve much of the rest of it. There are some organisms that eat oil, but they are only reclaiming the land at the rate of about 4 percent a year. It appears that these organisms work faster in some areas than in others as some areas were totally clean when they finally got around to dealing with cleaning up that particular area.

Charges were filed against Exxon in Exxon versus Baker, and the judge awarded $287 million for actual damages and $5 billion for punitive damages. Through a number of appeals, the punitive damages were reduced to $507.5 million dollars.

There were a number of reasons for the crash; among them was the lack of the proper operating equipment to detect obstacles, a tired crew, and a captain who was not on the bridge.

Of course, the crash of the Exxon-Valdez had nothing to do with global warming.

Arctic National Wildlife Refuge

Another area that offers great promise for oil development is the Arctic National Wildlife Refuge on the North Slope. Unfortunately, the propaganda against drilling shows pictures of pristine mountain areas far from the area that would actually be drilled. There was also concern for the wildlife, but notably wildlife congregate around drilling sites and the Alaska Pipeline probably because it is warmer there. Look this up on the Internet, and see the maps of where drilling will take place.

CHAPTER 41

The Pause in Temperature Rise

The pause stands for the fact that earth has not warmed statistically over the last twenty years. Many scientists have listed one year or another as being the hottest on record, but the amount warmer is actually less than the margin of error, so it is hard to make the claim. However, some have claimed that because temperatures vary considerably, it is unrealistic to say that one year is hotter than another or that several years represent a pause in temperature rise. However, decades without warming were not exceptional.

One of the problems you run onto is that there are several temperature datasets, and one might show general warming, and another might not show it. So it is difficult to know what to believe.

Throughout this whole discussion, it is important to remember that fifty years ago we believed we were going to have to take decisive action to avoid another ice age. In fact, looking at the history of ice ages, it appears we are about due for another ice age, probably in the next thousand years. So it really depends on to whom you talk, some suggesting the Earth is continuing to warm and others suggesting that the warming may have stopped. It is also important to note where your beginning and end points happen to be. If you start with a low point and end with a high point, you will show warming, and if you start with a high point and end with a low one, you will show cooling.

In January 2013, Dr. James Hansen and colleagues published their updated analysis that temperatures had continued at a high level despite strong La Niña conditions and said the "five-year mean global temperature has been flat for a decade, which we interpret as a combination of natural variability and a slowdown in the growth rate of the net climate forcing," noting "that the ten warmest years in the record all occurred since 1998." So what are you to believe: Is the Earth warming, or is the temperature flat? Dr. Hansen still believes that great danger lies ahead. He believes this is the last chance to save the human race from dramatic warming.

Over the next ten years, it will be interesting to see what the actual temperatures happen to be.

No matter what, some climate scientists have determined we have had less warming than the models have predicted, as though the many models were that reliable in the first place.

CHAPTER 42

Polar Bears: Numbers Increasing, Not Threatened

Introduction

What do you think of when someone mentions polar bears? Usually, two scenes come to mind: a mother with a couple of young playing in the snow or one bear clinging onto an iceberg about to fall into the ocean.

During the twentieth century, polar bear numbers were estimated to be at about five thousand. Today, they are closer to 25,000. Polar bear numbers in 1960 evidently were determined by talking to natives. Today (2019), they use helicopters, yet it is still very difficult to come up with hard numbers.

Secretary of the Interior Dirk Kempthorne listed polar bears as threatened on May 14, 2008, under the US Endangered Species Act because of what was considered a rapid decline in Arctic sea ice. However, today is ten years later, and their numbers have not declined.

Polar bears have survived several thousand years of Earth history in which the Earth warmed and cooled. To survive low ice years, polar bears would have had to hunt in the water or on land.

In 2007, scientists predicted that most of the bears would be gone in just a few years, yet they have been wrong, as today, the bear numbers are just as healthy as ever.

Diet of polar bears

Polar bears are almost exclusively carnivores, eating only meat. They will eat other things when nothing else is available, even garbage. Their main food is the ringed seal, and they usually catch them when they come up through their blowholes in the ice for air. If there is no ice, then they must be hunted in open water. The bears wait very patiently along the shore until the ice begins to melt, and they can start hunting.

Churchill, Manitoba, is a favorite haunt of the bears, being on the Hudson Bay. The bears come down to the bay to wait for the ice to melt. It is also a favorite spot for tourists as it is one of the few far north towns with a road and hotels. The town has very strict rules to keep people and bears apart, but tourists love to get pictures of the bears and sometimes get too close. (See *Eaten: A Novel* by Susan Crockford.)

While polar bears generally eat seals, the males have eaten their own offspring from time to time or even other adult polar bears (cannibalism). Evidence suggests that these were not always starving bears but healthy ones. Some polar bears starve even when sea ice is optimum. This is true for those that are still young (two to three years) and have not learned aggressive behavior in hunting. It is also true for the very old who can no longer hunt effectively.

Those who kill polar bears often think to eat them, and when the flesh is pretty tough, they settle on the liver. Don't do it. Polar bear livers often contain toxic levels of vitamin A.

Starving polar bears

There have been a couple of pictures of very emaciated polar bears with the caption that they were starving. Yet this seems very unlikely. Like any animal, polar bears get old and eventually die, and some polar bears get sick. If the pictures are used to get help for a subpopulation of bears, that would be great, but if they are taken totally out of context to try to convince people of a particular political position concerning global warming, that would be wrong. There are plenty of pictures of very healthy polar bears from all over the Arctic.

Swimmers

Scientists reported that four bears drowned while swimming, but it was later decided that they had drowned in a storm and were unable to make it back to land. These were noted while the scientists were flying looking for other animals, and it was not documented, so we really do not know for sure why they were dead. Polar bears have been known to swim great distances, swimming sometimes as far as sixty miles from shore.

Some of the bears swimming in the southern Beaufort Sea traveled up to four hundred miles. Many bears had radio collars and were clocked at over one hundred miles. Although no one knows for sure, some scientists still worry that declining sea ice may be causing bear cubs to drown, but other scientists believe that polar bear cubs can also swim great distances.

No bears are drowning on a regular basis in the Arctic.

I am told that bears can swim at 6.2 miles per hour; a really good swimmer (man) can swim only 5.3 miles per hour, so don't try to outswim a polar bear.

Bears have a lot of fat which keeps them warm in very cold water.

Polar bear bodies

Polar bears are the largest land carnivore in the world. When we see them romping with their young, we are easily misled, because they are actually very dangerous. Here are some facts about polar bears.

Male bears weigh between 770 and 1,500 lbs. (An average horse weighs 1,100 lbs. Horses can weigh as little as 200 lbs such as a pony and draft horses can weigh 2,000 lbs, or even 2,200 lbs, but remember they are not carnivores; they do not eat animals or people like polar bears do.)

A male polar bear stands between 7.9 ft and 9.8 feet. (An average grizzly bear is between six and nine feet standing and one measured about twelve feet.)

A female polar bear weighs 330–550 lbs and stands 5.9–7.9 feet.

A female bear is called a sow, a male a boar, and the young are called cubs.

Bears give birth in the winter when they are in hibernation, and they usually give birth to two, though one to four is common. The cubs are born between November and January. Cubs usually weigh between 22 and 33 lbs when they exit the den. A female gives birth to young only once every three years or so.

A female is able to bear young after four to five years, and a male is mature after six years.

There is only one species of polar bear, *Ursus maritimus*, which means maritime bear.

The hair on a polar bear is hollow, which provides excellent insulation, but it is still too cold for them in Antarctica. The bear has hair on its large feet which have five toes and long claws.

Location of Polar Bears

Polar bears are found all along the Arctic Circle, wherever there is land. There are several subpopulations, but they are all the same kind of bears.

Threatened

Polar bear numbers seemed to have declined because of unregulated hunting by natives. In 1973, because of overharvesting of bears, an international agreement was signed limiting the number of bears that could be killed. In 2008, polar bears were declared threatened by the Secretary of the Interior, Dirk Kempthorne. Kempthorne insisted that the bears were threatened because of projected loss of sea ice, not because of oil and gas development. It is important to note that the number of bears had not actually declined, but they thought they would decline because of declining sea ice. Of course, polar bears have survived for thousands of years through years of little or much sea ice, so it seems they will survive now, if we use a little common sense in their management.

Because polar bear numbers have risen and declined over the years, it seems obvious that their numbers are not controlled by world temperatures or by carbon dioxide levels. It is especially important to note that their numbers are not controlled by the small amount of carbon dioxide contributed to the atmosphere by humans.

Senator Ted Stevens said the polar bear numbers had tripled, but he was unable to document that number. Investigation concluded, however, that polar bear numbers had increased.

For an excellent article on some of the challenges of counting bears, check out the following article.[32]

Denning on ice or on land

Normally, polar bears make their dens in the fall in drifted ice or snow after the weather turns cold, and there is plenty of ice and snow. At times, they have had to den on land. It is during hibernation that the tiny cubs are born, in March or April. It takes about eight months for the cubs to be born.

Newborn polar bears weigh about twenty ounces, the size of a small cat. They are born blind (like a cat or rabbit). In eight months, they will weigh one hundred pounds. The mothers having hiber-

---

[32] https://www.factcheck.org/2008/06/polar-bear-population/

nated for several months and having borne several cubs, they have a tremendous need for food, usually the ringed seals. To catch the seals, they often congregate near Churchill, Manitoba, waiting for the pack ice to melt.

Killing people

Most people have enough good sense to stay away from polar bears, but there have been a few people killed by bears, very few.

There are good reasons to fear polar bears.[33]

Mating with grizzlies

Many scientists believe that polar bears and grizzlies do not mate, but there have been several examples in recent history to prove that they do. (It is possible that these mating pairs took place in a zoo.) As the Earth has warmed, it has enabled grizzlies to roam farther north into the territory of the polar bears. While polar bears usually breed on ice, grizzlies prefer land.

If the father is a polar bear but the mother is a grizzly, the offspring is called a "pizzly bear," and if the mother is a polar bear and the father is a grizzly, then it is called a "grolar bear."

The ability of polar bears to mate with grizzlies is just further proof that they are not really endangered. Scientists believed the polar bears were endangered, not because their numbers were declining, which they were not, but because they thought the numbers would decline in the future as the sea ice declined. It appears, however, that the bears have found food on land or learned to hunt seals in open water.

It is evident that polar bears are not endangered in any sense of the word, and further, it is clear that man has not caused their numbers to decline due to rising temperatures, because quite simply, their numbers are actually increasing.

---

[33] http://www.heavemedia.com/2013/02/04/reasons-to-fear-the-polar-bear/

CHAPTER 43

Pollution: Air and Water

Water pollution

Pollution was very common just one hundred years ago. And even today, it is common for some people to throw dead animals into the river just to get rid of them. We saw that a lot when we lived in Mexico. We took them out of the water; they put them back to wash out to the ocean, they said.

There are many types of pollution:

Pathogens can be bacteria, protozoa, or viruses. Bacteria are fairly common in water but do not become a problem until they begin to multiply. The most common are coliforms and *E. coli* bacteria, commonly found in human waste. When *E. coli* bacteria is found, it usually indicates that the water has been polluted by human or animal waste.

Some *inorganic material* is also found in water as a pollutant. This can include heavy metals such as arsenic, mercury, copper, chromium, lead, zinc, and barium. Though these can be harmless in very small quantities, in higher quantities, they can cause serious health problems and even death.

Organic materials are molecules with carbon in their makeup. Although some carbon compounds are necessary for life, water contaminated with some of these can cause serious illnesses.

There are also larger pollutants called *macroscopic pollutants.* The most common found is trash, especially plastic waste, which usually does not decompose. Steps have been taken to keep these out of the water, including the use of county trash dumps. In some third-world countries, this trash is just thrown out and the rain washes it into the rivers, creating pollution. Modern cities have programs to pick up trash, including plastic trash, and dispose of it in trash disposal sites. Someone has designed a plastic that does decompose, so it would be great for plastic bags. Most of the plastic in the ocean is not from America (we seem to know better) but from the East.

Wood, shipping containers, and even shipwrecks cause pollution; however, these are not very common. Any pollutant can affect aquatic life. So it is important to keep them out of the water. [34]

Water covers well over half of the Earth (70 percent) and is vital for all people and for all living organisms. Most common causes of water pollution are untreated household waste as well as industrial and agricultural waste. Some of this waste is a by-product of an industrial process, and in time, we have learned to prevent this pollution. Some processes have found that what was once pollution is now a valuable product. Oil spills are also a cause of pollution, but this is fairly uncommon. Not only is this pollution but loss of a very valuable product. Oil, however, eventually breaks down from the action of certain bacteria. Most household waste is now treated at sewage treatment plants or household disposal systems. Some of these systems treat water until it is almost clean enough to drink.

Ozone is also used to treat sewage. Generators use ultraviolet radiation or electricity to convert oxygen to ozone (O_3) which then oxidizes bacteria, mold, and organic compounds found in waste (and kills it).

---

[34] www.environmentalpollution.in/water-pollution/harmful-effects-of-water-pollution-on-man-and-aquatic-life/140

Septic tanks treat sewage by separating solids from liquids. Septic tanks rely on biological processes to degrade the solids, while the liquids flow out into a land drainage system, called an oxidation field. These are variously called leach fields, drain fields, and absorption fields. They are fairly close to the surface to allow oxygen to work on the liquid. Some places have oxidation or pickle ponds which serve the same purpose.

As our civilization has become more advanced, we have learned to value clean water more, and each of us takes steps to protect the water we must drink.

Water from waste

Today there are many processes where waste is purified into pure water. Some are fairly expensive, while others are fairly simple. Some even use a system of filtration to clean the water that is polluted. When we worked for the Navy on the desert in California, I noticed tomatoes growing in the lawns. I was told that tomato seeds can go through the septic process and survive to later grow tomato plants.

Desalination is a fairly simple process where the seawater is simply distilled. It is a fairly expensive process, but countries that have nuclear energy have found it a necessary solution. The largest desalination plant in the world is in Saudi Arabia which can produce 273 million gallons of drinking water a day, and in Israel, a quarter of the country's drinking water is generated by desalination.

There are other systems for providing clean water including harvesting fog and even using bicycles. Solar stills are also used, and while camping, we often built a solar still to collect a couple cups of water a day. It is a fairly simple process with a piece of plastic, a cup, and a hole in the ground. Pour the dirty water on the ground in the hole around the cup; place the plastic over the hole with a small rock in the middle, and then the water collects on the plastic and drips into the cup. The sun evaporates the water. It collects on the plastic and drips into the cup.

There are a number of ways to collect rainwater if you lack enough water to drink.

Air pollution

Not only is the water sometimes polluted, but the air is often polluted as well. Most of this pollution comes from burning different things. Most people know when they are burning something that pollutes the air.

Dust is sometimes considered pollution, but dust will settle and causes no long-term effects.

A lot of pictures have shown industrial plants or even electrical generation smoke stacks with a lot of pollution going into the air. Unfortunately (or fortunately), these pictures were taken when the sun was on the other side, so it made the cloud look like terrible pollution. It is, in fact, nothing but steam, like what comes out of your car exhaust when you first start it in the morning. (If the sun were on the other side, the vapor would look white.) The rest that comes out of the stack is nothing but carbon dioxide, which is a colorless gas and vital for all life. None of the things we have talked about above are caused by global warming. Plants depend on carbon dioxide, especially trees, and carbon dioxide is not a pollutant but a vital gas for all life.

CHAPTER 44

Portland, Oregon, and Renewable Energy

The city of Portland mayor Ted Wheeler, Mayor of the Oregon City of Portland, called the community into action in his statement on the commitment: "We don't succeed addressing climate change by government action alone. We need our whole community: government, businesses, organizations and households to work together to make a just transition to a 100 percent renewable future." He went on to note that he is fully aware of how challenging it will be: "Getting our community to 100 percent renewable energy is a big goal."[35]

The city of Portland is pretty excited about their new program. They have listed several milestones expected:

1. Phase out coal by 2035
2. Phase out natural gas by 2035
3. 100 percent renewable electricity by 2035
4. 100 percent renewable energy by 2050

(They are fairly safe making these predictions as many of the people who are alive now will be gone by 2035.)

It is hard to imagine our modern heavy industries without coal and natural gas, especially steel mills. Coal, oil, and natural gas are

---

[35] http://www.redgreenandblue.org/2017/05/26/portland-oregon-commits-100-renewable-electricity-18-years/

concentrated energy. To get the same amount of energy from wind and solar would take a lot of equipment.

No one has said what they plan to do when the wind is not blowing or the sun is not shining. Most now rely on natural gas to provide for electrical energy when it is dark or the wind is not blowing. What will they do when they cannot use these concentrated energy sources? They could use huge batteries or pump to a lake above (hydroelectric), which are both pretty inefficient, but they have not said.

It is extremely difficult to imagine making steel with just solar energy, when a lot of gas and coal are needed. One has to wonder how carefully they have thought through their plan. It looks like a lot of people will suffer before they give up on wind and solar, although they do have some hydropower from dams. People could even starve, which is the result when government seems to know more than science.

Portland has a lot of very smart people who should be able to see the problems with renewable energy. Can you imagine people bringing groceries to your favorite grocery store on bicycles or with small electric cars? To have electric trucks would require much of the truck be taken up with batteries, and even that would not work for long hauls across the nation. What about ships? Do we have to go back to sailing vessels?

Renewable energy could be used for some things, like lighting, but for much beyond that, it is pretty impractical. Even with all this renewable energy, it will not affect the production of carbon dioxide because much of this has a natural source.

CHAPTER 45

Sea Level Rise

The sea has risen at a more or less even rate since the end of the last Ice Age, about ten thousand years ago, or more specifically the Little Ice Age, which ended about 1850. During the last Ice Age, there was ice a mile thick where we live in North America (Idaho), and the sea level was down about 370 ft. Now, all that ice has melted, and the sea level is more or less stable because there is no more ice to melt, except on Antarctica, Greenland, and a few mountain glaciers. Right now, it is rising at a rate of about one-eighth inch a year or about seven to twelve inches a century.

Those who believe in a climate calamity want you to believe that the sea will rise and drown all the cities along the coast. The greatest concern seems to be for the Maldives, a group of small atolls and islands in the Indian Ocean. Al Gore claimed they were being drowned with seawater, but a London Court said his claims in *An Inconvenient Truth* were baseless. The judge said there was no evidence that any of them had left their island homes.

The alarmists want you to believe that the action in the movie *The Day After Tomorrow* is real, but though it is great fiction, there is no basis in fact, especially for a boat on a New York street.

The introduction of *Climate Change Reconsidered* on sea level has this to say:

> Sea-level rise is one of the most feared impacts of any future global warming, but public discussion of the problem is beset by poor data, misleading analysis, an overreliance on computer model projections, and a failure to distinguish between global and local sea-level change—all of which has led to unnecessary and unjustified alarm. [36]

Sea level is affected by many things including melting ice on Greenland and Antarctica and mountain glaciers. Local sea level can also be affected by wind currents. Also, water expands as it warms, and the oceans are very deep, so it takes thousands of years for them to completely warm, and as they warm, the level rises. The oceans also hold fifty times as much heat as the atmosphere, so it takes a long time to heat that much water. At the present rate of rise, retaining walls can be built to protect the land as the sea rises. (Think of Holland and their windmills.) In some locations, the land has subsided, so this gives a false sense of sea level rise.

Much has been in the news about the acceleration of sea level rise, but multiple tide gauges and satellite measurements have confirmed that the sea is continuing to rise at the same rate it has risen for many centuries. Some programs, such as satellite measurements, have actually registered a slowing of sea level rise.

Some climate scientists have claimed that the sea level will rise dramatically near the end of the century. Again, quoting from Climate Change Reconsidered:

> Hansen and Sato (2011) claimed sea level will rise 5 meters (16.25 feet) by 2100 AD and proposed an exponential increase in glacier melting would produce a 4 meter rise (13 feet) in sea

---

[36] *Climate Change Reconsidered II: Physical Science*, p. 756

level in the 20-year period from 2080 to 2100, a rate of 200 mm/year. These frenetic (rapid and uncontrolled) estimates are made notwithstanding that the Antarctic ice cap has expanded, not decreased, in the past 20 years and the Greenland ice cap was about the same size as today during the 2.5°C warmer Holocene Climate Optimum. [37]

Keeping this in mind, it is very unlikely that the oceans will rise faster than they are rising now, and there will be no danger that cannot be handled with proper planning. It seems very unlikely that global warming is causing the seas to rise faster than in recent history. Here is the conclusion of the Climate Change Reconsidered:

> In its 2007 report, the IPCC (United Nations Intergovernmental Panel on Climate Change) projected global sea level was likely to rise somewhere between 18 and 59 cm by 2100. Since then, several model-based analyses have predicted much higher sea-level rise for the twenty-first century, even exceeding 1 meter in some cases.
>
> In contrast, multiple careful analyses of tide gauge records by different research teams made it clear the acceleration of sea-level rise proposed by the IPCC and its scientists does not exist. Most records show either a steady state of rise or a deceleration during the twentieth century, both for individual records and for globally averaged datasets. In addition, and though only about 20 years long, the satellite radar altimeter dataset also records a recent decelerating rate of rise. [38]

---

[37] 5,000 to 9,000 years ago
    Ibid. p. 758
[38] Ibid. p. 775

If the Arctic were to melt, the sea would not rise at all as the ice there is merely floating on the sea. But if the mountain glaciers and Greenland and Antarctica were to melt, the sea would rise even more. The rate at which the Arctic melts varies from year to year; some years melting more and other years less. Greenland melts some, and glaciers fall (calve) into the sea, but Greenland has mountains and valleys, so these glaciers are not moving very much. In fact, one experience has shown that the snow in Greenland is getting deeper and deeper.

During World War II, a number of airplanes ran out of fuel and landed on the ice in Greenland. Evidently, the men were rescued, but the planes were not saved. In 1992, an expedition was undertaken to find one of the planes. Ultimately, using a new system of ground-penetrating radar, they found one of the P-38 airplanes.

> One of those P-38s, now known as Glacier Girl, was extracted from the ice in 1992 by the Greenland Expedition Society, 50 years after it had been reclaimed by the Earth. Pat Epps and Richard Taylor of Atlanta had been searching for the lost planes for years, finally discovering them in 1988 thanks to new radar tech of the time. The planes had been carried two miles from their original location, and by the time they were found, they were under 264 feet of solid ice. [39]

It was quite a process to rescue the plane, which included boring a large hole down to where it was. The plane was brought up in pieces and restored. See footnote 38 link. Look it up you can watch it fly again.

The whole point of the P-38 is that it was there for over fifty years, and nearly three hundred feet of ice had covered it. So it proves that Greenland is not melting but gaining new ice each year.

---

[39] https://www.popularmechanics.com/flight/a18943/glacier-girl-p-38-fighter/

Since the sea is rising one-eighth inch a year, this is not a problem for anyone. Even if it should amount to over a foot at the end of the century, measures could be taken to protect the land.

Only a few places in Antarctica are melting and those are far from the South Pole and a few places along the shore. The rest of Antarctica is gaining snow but not very rapidly, as Antarctica is a virtual desert where it snows very little.

So the rising of the sea is not caused by man-made global warming, simply because it was rising at about the same rate before man began burning fossil fuels. It is more likely that the little warming is caused by changes in the energy of the sun.

CHAPTER 46

Solar Power

There has been a lot of talk about solar electricity lately. Some insist that we can be 100 percent solar by a certain date, usually 2030 or 2050.

There are basically two kinds of solar: One type has large arrays of mirrors that concentrate heat/light on a boiler which causes the water to turn to steam and drives a generator. One of the largest is located at Ivanpah, California, on the desert.

Ivanpah is a small village on the desert not far from the Nevada border, near Needles. This is the largest solar thermal power plant in the world. As a young desert rat, we spent a lot of time at Ivanpah buying the few things we needed. (I doubt the town is still there.) We were working on our mine which is in the New York Mountains near there. The plant also uses natural gas at night to provide electricity when the sun is not shining. One of the problems is the number of birds that have been scorched by the heat from the mirrors. (We are told that cats kill millions of birds, but while cats kill sparrows, these solar installations, as well as wind installations, kill eagles and other raptors.)

The other type of solar electrical generation converts sunlight directly to electricity which then inverts (changes) it to AC electricity. Here is just one example of that type of installation.

Solar panels are also installed on commercial buildings and homes.

The biggest difficulty with all solar installations is that they must have an alternate source of power when the sun is not shining or it is cloudy. Usually, this is natural gas, but the gas generators have to be kept idling so they will be ready when needed. They use almost as much gas when idling as when running at full power, and this is a tremendous waste.

All solar installations produce electricity that is many times more expensive than our regular hydroelectric-, coal-, nuclear-, or gas-produced electricity. The industry is heavily subsidized, and it is doubtful it would exist without those subsidies.

One example of the solar industry was Solyndra which received a federal loan of 535 million. Ultimately, it was not successful and declared bankruptcy. What a waste.

Solar panels sound like the real answer to green energy, but they simply cannot take the place of hydroelectric, coal, nuclear, and natural gas. The only place they make any sense is when the end user is far from an electrical grid. We used panels with batteries when in Mexico, but it only provided enough electricity for our lights. To run the washing machine, for example, we had to fire up the gas generator.

Neither solar nor wind will ever reach the point where they are as efficient as natural gas. It is these other sources (natural gas, coal, nuclear, oil) that have made our country great, and wind and solar will probably never be able to compete on their own.

CHAPTER 47

Wind Energy

Wind has been used for several hundred years. Probably the first use of wind was to pump water.

They were very important to the Dutch when they were used to pump out water so they could use the land below sea level for farming. They were also used to power mills for making flour.

Sailing vessels have used wind for thousands of years.

Today's wind turbines are much more sophisticated (and expensive).

Wind energy has been used to make electricity for many years, and the benefit is that you do not have to pay for the wind; it just blows. However, there are many other costs involved—e.g, extra transmission lines, a 100 percent auxiliary power source needed, etc. Additionally, there are times when the wind does not blow, or when it blows too hard and the wind turbine has to shut down. Drive by an installation of wind turbines and you will see that several of them are not turning. They have to be installed far apart as each wind turbine interrupts the wind and limits the wind for the next wind turbine. Also, if they try to turn too fast because the wind is blowing hard, they have to shut down.

Whenever a wind turbine is not turning, or not providing maximum enough power to meet the current load demand, then the electricity must be backed up with other power sources, such as coal, natural gas, nuclear, or hydroelectricity. These machines must be running on idle so they are ready to take the load in a moment of time, and this uses almost as much energy as if they were the only power source.

Wind-powered electricity could be used to pump water to a reservoir on a mountain which could then let the water flow downhill to a generator, again producing electricity, but this process is very inefficient. A lot of power is lost in pumping the water.

Wind electricity could also be stored in huge batteries, but these are extremely expensive, if you have enough battery for a city.

The towers for modern wind turbines are between 300 and 400 feet tall, and the blades can reach to over 600 feet. The foundations are also up to fifty feet into the ground, and so all together, they are fairly expensive. Think of all the material that goes into making the wind turbine, as well as the concrete in the ground.

Then the generator produces DC electricity at fairly low voltage which must be inverted and transformed to line voltage. Between generator, inverter, and transformer, there is a lot of expense, and a lot of energy lost.

A significant problem is what to do with the wind turbine when it wears out. Sometimes they catch on fire even with men on top working on them. Some have lost their lives repairing wind turbines. Sometimes wind turbines catch on fire and no local fire department has equipment to handle a fire 300 plus feet off the ground. Remember, they have a concrete foundation that goes deep into the ground. It is just not a simple process.

Finally, there is a problem with birds. For some reason, a bird cannot see a turning fan, and many of them are killed each year. There are numerous other issues with industrial wind turbines—including nearby property devaluation, reducing agricultural yields in the community, reducing number of visitors to tourist communities, multiple adverse eco-system impacts, etc. Additionally there are dozens of studies that have concluded that some nearby residents experience health issues due to the low level sounds generated by these industrial machines.

This does not seem like a reliable solution to any perceived problem of global warming. Communities need to think long and hard about these wind turbines; they may pay several million dollars in land rent and taxes, and end up with nothing substantial to show for it. Even the loss of bats alone, which eat insects, leave many insects free to destroy valuable crops.

This is not a technical, economic or environmental solution to any perceived problem of global warming.

CHAPTER 48

Wind Health Hazards

Several people have expressed concern about wind turbines and the effect they believe they are having on human health. Some folk who have lived near these "fans" have moved away because they believed the turbines were affecting their health.

The pollution they have mentioned are low-frequency sound, dirty electricity, and ground current which can each, along with shadow flicker, contribute to ill-health among those who live near wind turbines.

Most people who live near these turbines blame the effect on the noise generated by the turbines. Some scientists have also noted that pressure waves outside the range of human hearing may also have very unpleasant side effects.

These can include such things as depression, irritability, aggressiveness, sleep disorders, fatigue, chest pain, headaches, joint pain, nausea, dizziness, stress, heart palpitations, and some other symptoms. Some have labeled this as wind turbine syndrome.

Part of the problem is that these wind turbines produce very low frequency sounds which can travel long distances; the exact distance does not appear to be precisely known.

Dogs are not the only ones that can hear these long-wave, low-frequency sounds; they affect people as well. This can include nausea, disorientation, or vomiting.

Other forms of pollution are caused by the electromagnetism of the generators and the transformers. Most people who live near wind turbines are not affected at all, though some perfectly healthy people have complained of trouble concentrating.

It would be easy to condemn wind energy totally on the basis of these complaints, but at the very minimum, these turbines ought to be located far away from residences to minimize any health hazards.

Just a possibility of health issues might be a fairly good reason to prevent their use near cities, especially as a means of controlling global warming. It seems that wind turbines do not reduce carbon dioxide notably and do not prove that man is the chief cause of global warming.

CONCLUSION

CONCLUSION

So, you have read the book. Do you still believe that the Green New Deal and everything that goes with it is true? This book has presented part of the truth. You can either accept or reject it. The choice is yours.

If we follow the program of the climate alarmists, the chicken little folk who believe we have to spend trillions to stop an imaginary climate change, then we will not provide for a constructive tomorrow; we will just be part of the continuing problem.

There were some great principles in the founding of America, which allowed us to use our coal, oil and natural gas to make America the greatest nation in the world: A nation that was ready and able to help other peoples when they had dire needs from floods, earthquakes, fires, and other calamities. If the program of the Green New Deal becomes the law of the land we will never be able to help the needy again.

There is much more to climate change than was shared in this book. Some of that is contained in the appendices which follow. There is also a list of great books, each one of which has more of the truth about climate change.

Will you end up living in the dark, or will you take the time to inform yourself so that we can continue to live in the light? Climate alarmism is a fraud, and hopefully you have seen that by reading this short book. But there is much more to understand. I trust you will make the effort to do so. If you believe the information in this book is wrong then I would encourage you to do further research. There is lots of information on the internet; just search for climate change, climate alarmism, or climate skeptics.

No matter what, thanks for taking the time to read this short book. Please share it with your many friends.

APPENDIX 1

1941 Department of Agriculture Paper on Climate: Analysis

Historical Perspective on Climate Change

An analysis of the book *Climate and Man; 1941 Yearbook of Agriculture* published by the U. S. Department of Agriculture. (1248 pages, with Index).

INTRODUCTION

1941 was a great year, among the warmest in the twentieth century. In fact the temperature fell from about 1941 to about 1975. The perspective of this book is important because the authors were not running around trying to convince people that the Earth is in grave danger because of a rising level of carbon dioxide, but these experts in their fields were working on solutions that would allow plants to live where they would not naturally grow. This book of 1941 needs to be compared with any pseudo-scientific articles written today. Today we have a totally different perspective. Rather than trying to help the poor of the world with clean water and good crops we are trying to get them to use less energy, a cultural deathtrap. To truly understand what is happening with climate today we need a historical perspective, and the 1941 Yearbook gives us some of that perspective.

The book is refreshing in that it is just matter of fact, taking life as it comes. It has many articles by many professionals, but the main value of the book is the development of life in the year 1941. Floods, hurricanes, tornadoes, fires etc. All these are taken as a matter of fact and dealt with. The perspective is that these things have always occurred and will continue to occur without any serious consequences if proper precautions are taken. The main theme throughout the book is not how can we change the world as it is, but simply how can we make the most of our present situation.

CLIMATE AND MAN—A SUMMARY

By: Gove Hambidge (Principal Research Writer, Office of Information, Department of Agriculture.)

> Certain kinds of data about climate have been carefully collected for decades—figures on rainfall, for instance; but when the conservationist or the crop specialist tries to connect these figures with the behavior of plants, he finds that they do not work. A plant will thrive in one region with a certain amount of rainfall and fail miserably in another with the same amount. Rainfall does not tell the [whole] story. Further investigation shows why. It is not the amount of rainfall, that counts, but many things besides the amount of rainfall. It depends on the nature of the soil, the amount of wind, the sunshine and cloudiness, the humidity of the air, the temperature—above all, the rate of evaporation and transpiration, which are affected by these other factors. (p. 4)

Notice there is nothing said about carbon dioxide, or anthropogenic (man-caused) global warming. In fact, I doubt in 1941 they had even heard the term.

The author says that "The history of the Earth can be traced back some billion and a half years by geologists." He says further "Suppose we imagine this vast period as a single year of time. During practically the whole of that year, according to geologic evidence, the climate of the Earth was much more genial and uniform than it is today." (p. 7) He goes on to point out that the Earth had warm periods, and cold periods (ice ages).

He then relates climate to time and history:

> Human beings, then, have seen only the more violent moods of the Earth. They were not here during the immense stretches of time when it was comparatively quiet and peaceful. (Ibid)

Hambidge then talks about the nature of the polar ice caps.

> The key to the climatic difference between normal and revolutionary times, Russell (Richard Joel Russell, see pp 67–95) holds, is the existence of the polar ice cap. During all times, normal or revolutionary, *climate obeys the same physical laws*, set by the nature of the Earth as a rotating ball with inclined axis moving round the sun and surrounded by an atmosphere. But the difference between polar ice and no polar ice is very great over the whole earth. Yet the balance between the two conditions is delicate. "A rise of 2^0 F. in the temperature of the Earth now would be sufficient to clear polar seas of all ice." *We would then be living in a "normal" climate.* (p. 8) (Emphasis added)

Hambidge quotes Russell further:

> Russell finds no adequate evidence of recurring climatic cycles; there is no proof that short-time climatic changes, at least, are anything more than

matters of chance. Of the many theories that have been advanced to account for long-time changes, especially the occurrence of glacial epochs, he discusses several briefly—changes in the angle of the ecliptic (the inclination of the Earth's axis), precession of the equinoxes (a cycle occurring every 26,000 years), variations in the suns' radiant energy as indicated by sunspots, changes in the atmosphere that might affect the amount of radiant energy reaching the Earth, changes in the amount of carbon dioxide in the atmosphere, volcanic dust. In most cases, he concludes, the cause suggested is not adequate to produce the effect, or there are other serious objections to the theory. (p. 9)

CLIMATE CHANGE THROUGH THE AGES

By: Richard Joel Russell (Professor of Physical Geography, Louisiana State University, Collaborator, Soil Conservation Service)

Russell first talks about what is a normal climate as far as geologic history is concerned.

> Normal geologic periods were times of quiet between [geologic] revolutions during which normal climates prevailed. Rocks deposited at such times indicate a minimum of relief—unevenness in height—and few signs of crustal unrest. The paleontologic record is one of widespread ranges in both plant and animal distribution. The early Cenozoic was such a time. Plants closely allied to some of our warm-climate types were flourishing in places such as Greenland, Spitzbergen, and other lands in the high latitudes where their growth is impossible today. Even more strikingly uniform were the temperatures in all latitudes during most of the Paleozoic and Mesozoic eras. (p. 75)

The Author quotes C.E.P. Brooks concerning the growth of the ice caps.

> Brooks [Brooks, C.E.P., 1926 Climate Through the Ages; A study of the Climatic Factors and Their Variations] has shown that a surprisingly small variation in temperature causes a change from open to ice-capped polar seas. (p. 76)

Here is some historical perspective.

> The oceans of normal times were much warmer than those of today. In the absence of polar ice, from which cold water, heavy because of its temperature, now creeps to abysmal depths and accumulates, it is probable that bottom temperatures were considerably higher.
> The distribution of both plant and animal fossils bears out these generalizations. (p. 77)

The author mentions that toward the end of the Paleozoic glacial deposits occur in tropical India. (p. 80) He goes on to note that we are still in an ice age.

> That we have not fully emerged from an ice age is evident from the fact that complete melting of the Greenland and Antarctic ice caps would result in raising sea level another 100 to 160 feet. (p. 81)

In 1984, Salt Lake was rising and consideration was given to a number of alternatives, among them pumping some of the water elsewhere. Ultimately, because of a change in local climate, it was not found necessary.

While we emphasize how the climate has changed and the temperature has risen we need to keep in mind that we have only had good instruments to measure weather for about 150 years.

> Instrumental climatology, the precise, modern science based upon actual measured observations, dates only from the middle of the nineteenth century. The thermometer and barometer were invented two centuries earlier, but systematic observations were not taken. The earliest rainfall record, in the modern sense, started at Padua, northern Italy, in 1725. Sunspots have been recorded since 1749. At the beginning of the nineteenth century there were only 5 places in the United States and 12 in Europe where worth-while observations were being taken. The Challenger Expedition, 1872–76, brought back the first significant observations taken at sea. Even today [1941] the climatologist finds data too meager for satisfactory conclusions in nearly all parts of the world. (p. 84)

Of course today we have more land based weather stations, weather balloons, and satellites, (2019) but many weather stations have been moved from one location to another, usually from a grassy field to an airport, so the record is not consistent. Moving to an airport means setting on concrete which usually reflects a higher temperature, especially at night. This makes it look like the climate is increasing in temperature, when actually it is not, or not at the same rate.

By the end of the bronze age, about 1,000 B.C., the climate had changed.

> European climates were warm enough at that time to permit the growing of crops high in the mountains, in places now glaciated. A new period of flooding, starting about 850 B.C., drove the

people to drier and warmer localities. Similar histories have been traced for dwellers around lakes in northern Africa and the enclosed lakes of western and central Asia. (p. 87)

So what did the climate look like in the not so distant past?

From the climatic standpoint, warm and moist conditions lasted from about 5,000 to 3,000 B.C., and the time is called the Atlantic period. Temperatures were high enough so that all small mountain glaciers of the Alps and the present United States disappeared completely. (p. 89)

The author points out that climatic conditions at that time were not the same the world over, but varied from place to place. Some places were quite wet for a couple thousand years, then had severe drought conditions. Our recent droughts in parts of the Southwest, then, are not unusual in terms of the geologic/climatologic record. Some places widely separated were quite wet, then they turned dry.

About 400 B.C. a precipitation maximum is indicated in North America, Africa, western Asia, and Europe. All of these places record very dry conditions about A.D. 700. There is thus considerable evidence in favor of world-wide climate swings. On the other hand, the records indicate some notable exceptions, particularly between European and Chinese precipitation. (p. 90)

Worldwide climate swings were common in the past, and obviously not caused by the burning of fossil fuels by man.

He goes on to say more on the subject.

From about A.D. 180 to 350 Europe experienced a wet period. The fifth century was dry in Europe and western Asia and apparently also in North America. Many of the lakes in the western United States appear to have dried out completely. Europe was both warm and dry in the seventh century. Glaciers retreated to such an extent that a heavy traffic used Alpine passes now closed by ice. Tree rings in the western United States indicate minimum precipitation at this time. Nile floods were low until about A.D. 1000. (ibid)

If anything can be said about climate, the only thing that seems to be constant is that climate is always changing. Here in Idaho we say if you don't like the weather just wait and it will change. We are on the edge of competing air currents which means we can have at one time very arctic air, or just a few days later, sub-tropical weather.

The beginning of the ninth century brought heavier precipitation to Europe. The levels of lakes rose, and people living around their borders were pushed upslope. Documental evidence from south-western Asia and American tree rings give similar testimony. Warm, dry conditions returned during the tenth and eleventh centuries. This was a time of great exploratory activity among northwestern Europeans. The Arctic ice cap may have disappeared entirely. In any event the logs of Greenland voyagers show routes of travel where they would now be impossible because of ice floes. Greenland was settled in 984 and abandoned about 1410. (pp. 90, 91)

The author notes the difficulty of developing climate trends when we have only had good instrumentation for a few short years.

Though firm advocates of climate cycles will sharply disagree, such facts as we possess today

neither definitely demonstrate nor disprove the
existence of any real cycle. Such climate variability
as has been observed may be explained as resulting wholly from random fluctuations. (p. 92)

Random fluctuations: We call this just plain weather.

The author develops some theories based on the tilt of the axis
of the Earth toward the ecliptic, the plane on which it revolves around
the sun. The angle of the Earth in relation to the plane of the ecliptic
was 23°27'3" in 1941. It is estimated that it will reach a minimum of
22°30' in 9,600 years. (Present angle 2017, 23°26'13.3", and it varies
from 22.1 to 24.5 in a 41,000 year cycle.) This inclination of the axis
of the Earth is the principal cause of our seasons.

There is another factor, a cycle that happens every 26,000 years.
That is the fact that the orbit of the Earth is not completely round,
so at some points at various times of the year the Earth is either closer
or farther from the Sun.

Another factor that could affect climate is sunspots, variations
in the radiant energy from the sun, but in 1941 they did not know
much about the relationship between the two: (how changing sunspots affect weather). Today (2017) there is a much stronger case for
variations in solar energy affecting changes in climate.

Another thing that may affect climate is the atmosphere and the
relation to the Sun's energy reaching the Earth.

The ability of the sun's radiant energy to travel
from the outer limits of the atmosphere to the
Earth's surface may change with changes in the
atmosphere itself; on a clear day the coefficient
is higher than on a day with a heavy cloud layer.
Such variations might produce results similar
to those caused by variations in the emission of
radiant energy by the sun itself. (p. 93)

The author goes on to suggest why this might be a factor in a changing climate.

> If the atmosphere had a perpetual cloud layer, a great deal of solar radiation would be reflected back to space, and consequently the amount of energy available to the Earth's surface in the form of heat would be diminished. A cloud blanket, however, would also cut off a good deal of terrestrial radiation, tending to conserve such heat as might exist beneath it. Temperature ranges between day and night, one season and another, and higher and lower latitudes would be reduced. (p. 94)

So, which is more important, the reflection (from the clouds) of the energy from the Sun that comes to the Earth, or the insulating quality of the cloud cover?

The author also talks about carbon dioxide.

> Much has been written about varying amounts of carbon dioxide in the atmosphere as a possible cause of glacial periods. The theory received a fatal blow when it was realized that carbon dioxide is very selective as to the wave lengths of radiant energy it will absorb, filtering out only such waves as even very minute quantities of water vapor dispose of anyway. No probable increase in atmospheric carbon dioxide could materially affect either the amount of insolation [heat from the sun] reaching the surface or the amount of terrestrial radiation lost to space. (ibid)

But what about volcanoes? More dust in the air will affect the solar energy.

> Large amounts of volcanic dust in the atmosphere have also been considered as a possible cause of glacial climates. Lowered temperatures have followed great dust-producing volcanic explosions during the period of instrumental observation. Volcanoes have been particularly active during times of glacial climate. It seems most reasonable, however, to relate both the volcanic activity and the climate to crustal unrest [earthquakes and continental drift] and to regard the former more in the light of modifying influence than as the underlying cause of the latter. (ibid)

Volcanoes spew out large quantities of carbon dioxide and acid, which could also affect cloud formation as well as affect weather in other ways.

The conclusions of the author leave one with the feeling that it would be very difficult to predict what the climate will do in the future.

> Man has observed that climate conditions fluctuate rather widely from time to time at a given place, and in seeking to understand such natural phenomena he has been tempted to explain such fluctuations on the basis of recurring cycles. As yet, [1941] however, no definite proof has been advanced to contradict the opinion that all such relatively short-term climate changes are nothing more than matters of chance. The world pattern of climates today is the product of climatic variations, not the expression of recurring mean, or normal, conditions. The extent of desert climate will not be the same next year as this. The

humid margin of the desert is the product of an ever-changing distribution of extreme aridity. The time may come when such changes will be well enough understood to be of definite forecast and economic value, but it is likely that such information will be the fruit of long-continued and patient research. (p.95)

The rest of the book deals with the effect of climate on particular plants. The book is still available and it is suggested that anyone interested obtain a copy.

CONCLUSION

The book has dealt with the various factors that affect climate. What effect does the climate have on the various crops that are grown in the United States and elsewhere? Rather than looking at the various negative aspects of a particular climate the book seeks to mitigate those conditions, either in manipulation of different natural resources, or in the modification of the various plants to be able to survive when it is either too hot, too cold, too wet or too dry. Rather than doing as they did, we waste a lot of time trying to place the blame on someone or some group of people or industries. Carbon dioxide is probably not the chief cause of increases in global temperatures, but the Sun is the primary source of heat, and carbon dioxide is probably coming from the oceans and elsewhere. We also learn that though the atmosphere contains a very small percentage of carbon dioxide it is also the source of all living matter, both plant and animal. Simply put, plants take in carbon dioxide and give off oxygen. It is a simple life cycle that has gone on for thousands of years. It is up to us to make the most of it.

APPENDIX 2

Hansen/Hollingsworth Storms of My Grandchildren

Analysis of *Storms of My Grandchildren* by James Hansen, PhD [40]
A REVIEW OF JAMES HANSEN'S BOOK:
"STORMS OF MY GRANDCHILDREN..."
by Jim Hollingsworth | June 20, 2010
Reference: Hansen, James. *Storms of My Grandchildren: The Truth About the Coming Climate Catastrophe and Our Last Chance to Save Humanity.* Bloomsbury, USA, 2009.

INTRODUCTION

While scientists the world over continue to study and debate what part man has played (if any) in the gentle warming that took place mainly in the latter half of the last century, Dr. James Hansen is absolutely certain. The purpose of his book is to scare us into taking immediate and drastic action to control greenhouse gases, mainly carbon dioxide. Although Dr. Hansen is a scientist, his work is more political than scientific. He makes an emotional appeal and he does it by attempting to build fear that there will be no world left for our grandchildren to live in.

---

[40] http://www.edberry.com/SiteDocs/PDF/Hollingsworth_Review_Hansens_book.pdf

Few people are really concerned about global warming. Most know that change is the one constant you can expect with any climate. When doing surveys of what is important to people, global warming is generally listed dead last. Those who believe that our world is headed for disaster will welcome Dr. Hansen's book: He may also influence a few gullible. But, most of the rest of the world are more concerned about raising their standard of living, and even having things like clean drinking water and safe sanitary facilities, than they are about global warming. Having cheap and universally available energy, especially electricity, has saved many lives the world over.

Because of the fact that climate change is recognized as just a normal part of the history of the Earth most people are simply not convinced by the kinds of arguments put forth by Dr. Hansen and other global warming alarmists. The alarmists are like Chicken Little who ran around saying: "The sky is falling; I must go and tell the King". But there is no more substance in their predictions than there was in Chicken Little's.

Dr. Hansen's view depends on two things. First, that man, through burning fossil fuels (coal, petroleum, natural gas), is producing huge quantities of carbon dioxide, and this increase in carbon dioxide, through the "greenhouse effect", is causing the Earth to warm.

There is, however, considerable evidence that gentle warming and increased carbon dioxide are beneficial to plants. Also, more people die in winter than in summer. The present level of carbon dioxide is about 390 ppm (parts per million). [2019 carbon dioxide level 0.041 per cent, 410 parts per million (ppm)]. All plants need carbon dioxide; that is what plants use to grow: The more carbon dioxide the faster they grow. (Current level of carbon dioxide, Feb 2019 is 410 ppmv)

But, there is a remarkable similarity between the historical temperatures and the level of carbon dioxide. What we think we know about temperature in the distant past (before thermometers) comes from proxy data like tree rings, sedimentary deposits, and more recently, ice core samples, especially from Antarctica and Greenland. In Antarctica, the Russians have drilled down to a depth representing about 800,000 years. (The ice over Antarctica averages about 7,000 feet, and in some places is over 15,000 ft thick.)

Detailed analysis of these ice cores has revealed that temperature rises first followed about 800 years later by an increase in carbon dioxide. Do keep in mind that Antarctica is mostly a desert, with very little snowfall (less than one inch a year at the pole), so it is very difficult to separate one year from another.

The second thing that Dr. Hansen has put forth to support his case is what he calls "tipping points". Sure, he says, the Earth is just warming mildly at the present, but through a mechanism he calls "positive feedbacks" and "forcings" a little warming is going to be multiplied until a tipping point is reached and there is a dramatic rise in temperature, causing the destruction of all life on the Earth. The only problem with tipping points is that there is little historical evidence to support it. It is all based on computer models which are actually no better than the data that is put into them.

His idea about tipping points is no doubt based on the hockey stick graph developed by Michael E. Mann. This graph was popularized by Al Gore in his movie "An Inconvenient Truth" in which he mounts a man lift to reach the top of the graph. The only problem with the graph, of course, is that you have to exclude the Medieval Warm Period and the Little Ice Age and then project a short term tail of the graph out into the future to make it work. The graph has been discredited by Ross McKitrick and Stephen McIntyre and even the IPCC (Intergovernmental Panel on Climate Change) has stopped using it.

So, all things considered Dr. Hansen has no case. The science simply will not support any program requiring immediate and radical action to not only control, but also reduce, carbon dioxide in the atmosphere. He knows that only by appealing to our emotions and not to science will he gain any adherents to his program. Thus, the thrust of his book.

To read the entire comment check out the links below.
Storms of My Grandchildren (Analysis) by James Hansen
The first is in Word format.[41]
The second is a PDF file.[42]

---

[41] http://www.edberry.com/blog/climate-authors/jim-hollingsworth-climate-authors/review-of-hansens-storms-of-my-grandchildren/

[42] http://www.scienceandpublicpolicy.org/wp-content/uploads/2010/07/storms_of_my_grandchildren.pdf, www.edberry.com/SiteDocs/PDF/Hollingsworth_Review_Hansens_book.pdf

APPENDIX 3

My Testimony

Many people have wondered why so few Christians, especially Republican Christians, are concerned about the Earth burning up unless we spend billions on programs to stop the use of fossil fuels. I think it is because the Bible-believing Christians believe the promises that are in the Bible.

Probably the most important promise is the one God gave after the flood:

"And I will establish my covenant with you; neither shall all flesh be cut off any more by the waters of a flood; neither shall there any more be a flood to destroy the earth" (Genesis 9:11).

The alarmists have told us that the waves of the sea will cover parts of Florida and islands in the Pacific, but here is God's promise concerning the waves:

"And said, Hitherto shalt thou come but no further: and here shall thy proud waves be stayed?" (Job 38:11).

God also talks about the weather and the part it often has in His plan for the world:

"And I will lay it waste: it shall not be pruned, nor digged; but there shall come up briers and thorns: I will also command the clouds that they rain no rain upon it" (Isaiah 5:6).

God says that He would destroy earth with a fire, but it would be His decision and His action, not the action of man.

> The Lord is not slack concerning his promise, as some men count slackness; but is longsuffering to us-ward, not willing that any should perish, but that all should come to repentance. But the day of the Lord will come as a thief in the night; in the which the heavens shall pass away with a great noise, and the elements shall melt, with fervent heat, the earth also and the works that are therein shall be burned up. (2 Peter 3:9–10)

Every Christian knows that this Earth will be burned up, but it will be God's decision and in God's time. Until that time, everything will continue, and life will go on.

All this makes little sense if we have denied the one who died for us.

When speaking to Martha, who was concerned about her brother Lazarus who had just died, Jesus had this to say:

"Jesus said unto her, I am the resurrection, and the life: he that believeth in me, though he were dead, yet shall he live: And whosoever liveth and believeth in me shall never die" (John 11:25–26).

Living or dying our eternal life depends on our relationship with the one who died for us, the Lord Jesus Christ. It is going to make little difference on what we have done to save the Earth, but it will make a lot of difference as to our relation with the one who created the heaven and the Earth. It is a simple matter, and if you want to know more, drop me a note at jimhollingsworth@frontier.com, or just read the Bible (especially the Gospel of John or the Epistle to the Romans) where the answers are found.

APPENDIX 4

The Myth of Global Warming: Short Paper

This is a short presentation of global warming to share with your many friends.

Jim Hollingsworth
By Jim Hollingsworth
2009, Rev 5–9–17; Rev 2–1–19; Rev 2–26–19

INTRODUCTION

When Al Gore screams out in Senate testimony: "The Earth Has A Fever" he is not so much developing a scientific theory as he is seeking to use fear to drive a political agenda. The Intergovernmental

Panel on Climate Change (IPCC) has issued five (sixth in progress) reports and in each one they have become more convinced that man is the chief cause of global warming, and that this warming is and will be seriously destructive to life as we know it.

We are told that "the science is settled" and that there is "a consensus" of scientists who believe we are headed for disaster if we do not stop burning fossil fuels. Yet, there is a growing number of scientists who disagree. Over 32,000 scientists have signed "The Petition Project", over 9,000 of them with PhDs, proclaiming that man is not the chief cause of warming and that this warming will not be disastrous.

True science does not depend on a consensus, but on a careful analysis of evidence as found in nature. One scientist noted that once they have a theory about something they work hard to prove themselves wrong. History is replete with examples where scientists were just plain wrong about life matters, but continued research revealed the truth.

It is our contention that the present emphasis on man-caused (anthropogenic) global warming is a myth, in fact a carefully orchestrated hoax, not to further science, but to gain more control over the peoples of the world. It is incumbent on each of us to search out the truth in this matter and act accordingly.

THE EARTH IS WARMING

There seems to be little disagreement among scientists that the Earth is warming. In fact, the Earth has continued to warm since the end of the last Ice Age, about 10,000 to 20,000 years ago. But that warming has not been an even warming, but years of warming followed by some years of cooling, but each period leaving us just a little warmer than before.

We are told that the Earth has warmed about half a degree centigrade a century for the past 150 years. The actual amount of warming as recorded is difficult to support given that a half-degree is about all the closer we could record temperature until very recently. There are about 1,221 weather stations in America and few of these have been in the same position for the entire time. Some have moved to new locations, and others have had cities grow up around them, which raises the average temperature. Temperature readings can vary a couple of degrees depending on whether they are next to a building, on a slope, in a valley, or on a hilltop.

There appears to have been more warming in the Northern Hemisphere than in the Southern Hemisphere, but it needs to be kept in mind that there are few weather stations over the ocean, in Africa and South America, and even fewer in Antarctica. It is very probable that the Earth is warming, but by less than half a degree. (The Northern Hemisphere has more land than does the Southern Hemisphere.)

CARBON DIOXIDE IS A POLLUTANT

The U.S. Supreme Court has declared that carbon dioxide is a pollutant. That is not a scientific statement made by a scientific body, but a deliberative statement made by a political body. We often forget that Supreme Court Justices are human just like the rest of us, and though we expect them to know more about the law than we know we do not expect them to be experts in every field of endeavor. The greatest evidence for this is in their own decisions (or opinions) which are often 7–2, 6–3, or 5–4. One vote one way or the other would change the outcome of the decision.

Our atmosphere is approximately 78% nitrogen, 21% oxygen, .9% argon, and .04% carbon dioxide. From that "trace" amount of carbon dioxide is built all the plants we see on the Earth. Far from being a pollutant, carbon dioxide is a natural substance required for all plant life. All plants use carbon dioxide to grow and in the process they give off Oxygen. Animals use Oxygen and give off carbon dioxide. This relationship is the miracle of life that enables both plants and animals to survive on the Earth.

Growers know that increasing carbon dioxide increases plant growth, and for this reason carbon dioxide is sometimes used in greenhouses to increase plant growth. It has also been demonstrated that an increased level of carbon dioxide enables a plant to survive and even thrive at warmer temperatures.

Carbon dioxide is not a pollutant.

CARBON DIOXIDE IS A GREENHOUSE GAS

The principle of a greenhouse is quite simple. Light enters through the glass and strikes the surface. It is transformed into infrared rays which are longer and do not so easily pass back through the glass. You experience the greenhouse effect when you leave your car shut up in the summer and notice how hot it is.

The greenhouse effect we note for the Earth is a little more complicated. The infrared rays that are re-emitted from the Earth are actually trapped by the greenhouse gases, which warm the gas and heat the Earth. It is this effect which makes our Earth habitable; otherwise it would get very hot in the day time, and very cold at night; extremely hot in summer, and extremely cold in winter.

The main greenhouse gas is water vapor. Where there is more moisture in the air the climate is more tempered. Thus while daily temperatures may vary on the desert as much as fifty degrees they vary only a little in the tropics.

Water vapor is the most important greenhouse gas, and the next is carbon dioxide. However, doubling the amount of carbon dioxide in the atmosphere will not double the temperature rise. There is a definite limit, and that limit is determined not by the amount of greenhouse gases in the air, but by the amount of solar radiation "reflected" from the Earth. Once all the infrared rays have been "captured" by the greenhouse gases there can be no additional increase in temperature.

(Some scientists are convinced that the atmospheric greenhouse effect is not as simple as a domestic greenhouse.)

MAN IS NOT THE CHIEF PRODUCER OF CARBON DIOXIDE

There appears to be a definite link between temperature and carbon dioxide. Early analysis of ice cores seemed to indicate that as the carbon dioxide increased it caused a rise in temperature. Subsequent analysis of the core data has revealed that the temperature rise came first, followed about 800 years later by an increase in carbon dioxide. And some years, while the carbon dioxide was rising, the temperature of the Earth was not rising.

There is no question that man produces carbon dioxide. He produces carbon dioxide simply by breathing. But, he also produces carbon dioxide by burning fossil fuels. Every fossil fuel except Hydrogen produces carbon dioxide when it burns. Some products produce more carbon dioxide than others, depending on their chemical composition. Methane produces less carbon dioxide and wood produces more. It is important to keep in mind that when wood burns it produces carbon dioxide, but when that same tree dies and rots it also produces carbon dioxide.

The ocean is a tremendous storage tank (carbon sink) for carbon dioxide. But, as the oceans warm they can hold less carbon dioxide. The warming ocean releases carbon dioxide into the atmosphere. An increase in atmospheric carbon dioxide will cause more carbon dioxide to be absorbed by the ocean. But, which comes first is difficult to know. It is much more likely that some natural factor such as changing solar radiation is actually warming the Earth, warming the oceans, and thus increasing the carbon dioxide in the atmosphere.

RISING TEMPERATURES ARE NOT HARMFUL

We are told that there will be great species extinction because of rising temperatures. We pick up a magazine and there is a polar bear standing on a small iceberg hoping to survive. The truth is that most populations of polar bears are expanding and loss of sea ice is not a problem for them. (Polar bear populations have increased from about 5,000 in 1950 to the present of about 25,000.) Most of the loss of sea ice area has been caused by changes in wind patterns.

Rising temperatures are actually beneficial to most plants and animals. While humans can adapt from the very cold to the very warm, most people prefer a warmer climate, and many migrate to the south during the winter. Plants actually increase their habitat when temperatures warm, moving higher in latitude (farther north) and higher in elevation (up mountains) when conditions warm, but they still maintain their present habitat. Animals, of course, can easily move to cooler climates if they prefer. Actually plants do well with warmer temperatures especially when the rise in temperature is accompanied by a rising carbon dioxide level. Plants are affected more by rainfall than they are by temperature.

As far as people are concerned it is important to note that more people die each year from cold than die from the heat.

REMEDIATION VS ADAPTATION

Most of the measures suggested to remedy higher levels of carbon dioxide (Remediation) will be deadly for world populations, especially in underdeveloped countries. Because gasohol is produced from corn, efforts to substitute alcohol for gasoline is causing worldwide food shortages, and raising the price of most grains. This has become so critical that some places have even had riots over food prices. (Did you know that carbon dioxide is a byproduct in the production of alcohol? Are we saving more carbon dioxide by using gasohol than is produced in the production of that same product?)

The United States was not a signer of the Kyoto Protocol (treaty). But, those countries that did sign have greatly missed the mark. The only way to reduce carbon dioxide production is to reduce burning of fossil fuels: gasoline, diesel, coal, natural gas, wood etc. These are the main sources of energy, and though we might be able to cope by building more nuclear power plants, much of the developing world will be very dependent on these other resources over the next several decades.

And, even if we were able to reduce our carbon footprint, most of the rest of the world would not. China and India are two very large developing countries and have expressed no interest in cutting production of carbon dioxide. Unless they participate in remediation what we do will have very little total effect.

Rather than causing world poverty by reducing carbon dioxide it will make a lot more sense to adapt to a changing temperature.

DOES CATASTROPHE AWAIT INACTION?

Floods, drought, hurricanes, these, we are told, are all signs of coming catastrophe from global warming. Yet, these are normal parts of any climate. Climates change from warm to cold, from wet to dry. Our Earth has weathered many serious changes in climate, from the Ice Ages to ages of tremendous plant growth (with high levels of carbon dioxide) yet the biosphere (plants and animals) has survived. We shall continue to survive because we are very adaptable. And, as time goes on we will develop even more innovative ways for adapting to changing climate. Although Alarmists would have you believe otherwise, none of the predicted changes will happen suddenly and there will be time to adapt.

Some politicians know that they can gain more control over our lives through programs designed to remedy global warming. We need to be vigilant to see this does not happen. We can make better use of our energy; that just makes good sense and helps each one of us. But, most of the suggested programs from wind farms to solar energy to hybrid automobiles will cost almost as much energy to produce as they actually save.

Common sense demonstrates that this great planet has survived many changes in climate and will do so again. Government programs to help would be far better directed toward clean water the world over, as well as better sanitary conditions and food production. Even control of such things as malaria would be a greater help to mankind than any effort to control global warming. Besides, most of what you hear about global warming is a myth.

Jim Hollingsworth is a retired building contractor in Kootenai County, Idaho. He has run for State Representative three times and is active in causes of liberty in Idaho. He receives email at: jimhollingsworth@frontier.com

APPENDIX 5

The Father of Climatology Speaks

By Noel Sheppard | June 18, 2007 11:46 AM EDT

Reid Bryson, the 87-year-old considered to be the father of scientific climatology, has once again spoken out strongly against anthropogenic global warming theories being regularly disseminated by alarmists in the media and the scientific community.

In an interview published by Wisconsin's Capital Times Monday, Bryson spoke about the money involved in this "religion," and when asked about soon-to-be-Dr. Al Gore's schlockumentary "An Inconvenient Truth" marvelously responded (emphasis added throughout):

"Don't make me throw up…It is not science. It is not true."

But Bryson had loads more to say on this issue (better fasten your seatbelts!):

There is no question the earth has been warming. It is coming out of the "Little Ice Age," he said in an interview this week.

"However, **there is no credible evidence that it is due to mankind and carbon dioxide. We've been coming out of a Little Ice Age for 300 years. We have not been making very much carbon dioxide for 300 years. It's been warming up for a long time,"** Bryson said.

Think Bryson will be interviewed any time soon by Katie, Charlie, or Brian? Regardless, the article continued:

The Little Ice Age was driven by volcanic activity. That settled down so it is getting warmer, he said.

Humans are polluting the air and adding carbon dioxide to the atmosphere, but the effect is tiny, Bryson said. **"It's like there is an elephant charging in [the room] and you worry about the fact that there is a fly sitting on its head. It's just a total misplacement of emphasis,"** he said. **"It really isn't science because there's no really good scientific evidence."**

Just because almost all of the scientific community believes in man-made global warming proves absolutely nothing, Bryson said. **"Consensus doesn't prove anything, in science or anywhere else, except in democracy, maybe."**

For the alarmists who love to depict every skeptic as being on the take of oil companies:

Bryson, 87, was the founding chairman of the department of meteorology at UW-Madison and of the Institute for Environmental Studies, now known as the Gaylord Nelson Institute for Environmental Studies. He retired in 1985, but has gone into the office almost every day since. **He does it without pay.**

"I have now worked for zero dollars since I retired, long enough that I have paid back the people of Wisconsin every cent they paid me to give me a wonderful, wonderful career. So we are even now. And I feel good about that," said Bryson.

How refreshing. Can the alarmists in the scientific community—and folks like soon-to-be-Dr. Al Gore—claim that they are doing their "work" for nothing? Read on:

So, if global warming isn't such a burning issue, why are thousands of scientists so concerned about it?

"Why are so many thousands not concerned about it?" Bryson shot back.

"There is a lot of money to be made in this," he added. "If you want to be an eminent scientist you have to have a lot of grad students and a lot of grants. **You can't get grants unless you say, 'Oh global warming, yes, yes, carbon dioxide.'"**

Bryson then pointed out how the media work:

Reporters will often call the meteorology building seeking the opinion of a scientist and some beginning graduate student will pick up the phone and say he or she is a meteorologist, Bryson said. **"And that goes in the paper as 'scientists say.'"**

The word of this young graduate student then trumps the views of someone like Bryson, who has been working in the field for more than 50 years, he said. "It is sort of a smear."

Of course, then some ignoramus will copy the opinions of this graduate student and disseminate them throughout the blogosphere as yet another example of the consensus. NewsBusters members should be very familiar with this:

Bryson said he recently wrote something on the subject **and two graduate students told him he was wrong, citing research done by one of their professors. That professor, Bryson noted, is probably the student of one of his students.**

"Well, that professor happened to be wrong," he said.

Sound familiar? And this is why many view all of this nonsense as "junk science":

"There is very little truth to what is being said and an awful lot of religion. It's almost a religion. Where you have to believe in anthropogenic (or man-made) global warming or else you are nuts."

Yep. I've witnessed that as well. Of course, what is the real science involved:

While Bryson doesn't think that global warming is man-made, he said there is some evidence of an effect from mankind, but not an effect of carbon dioxide.

For example, **in Wisconsin in the last 100 years the biggest heating has been around Madison, Milwaukee and in the Southeast, where the cities are. There was a slight change in the Green Bay area, he said. The rest of the state shows no warming at all. "The growth of cities makes it hotter, but that was true back in the 1930s, too," Bryson said. "Big cities were hotter than the surrounding countryside because you concentrate the traffic and you concentrate the home heating. And you modify the surface, you pave a lot of it."**

Yet, the most wonderful part of this interview was:

Bryson didn't see Al Gore's movie about global warming, "An Inconvenient Truth."

"Don't make me throw up," he said. "It is not science. It is not true."

Bravo, Mr. Bryson. And thanks.

It has to be obvious from the above interview that man has little to do with global warming.

APPENDIX 6

Paradise Fire in California

Trouble in Paradise: Why This Fire Should
Not Have Taken Place

By Dr. Bob Zybach, PhD

I have been a co-moderator of, and irregular contributor to, a discussion blog titled "Global Warming Realists" for several years. We have more than a hundred members, including a number of scientists on at least three continents, professional meteorologists, knowledgeable citizens, a few politicians, and at least one member of the media. Lars Larson—who hosts a popular radio talk show that exists on two levels; his afternoon "Northwest Show" that covers Oregon, Washington, Idaho, and northern California, and his evening "National Show," which also covers most of the remaining 46 states and much of southern Canada.

In early November, as part of a blog discussion regarding the Paradise Fire in northern California, I wrote:

"The area around Paradise had been oak savanna maintained by annual Indian fires for many centuries. Then the Spanish and their cattle moved in and the area was kept free of flash fires by mass seasonal grazing for another century or more. Then the USFS (US Forest Service) banned cattle grazing, too, and predictable wildfires have followed.

The reason the towns burn so quickly is that they are constructed of kiln-dried lumber—essentially a bone-dry forest dismembered and moved onto a prairie.

This has zero to do with "climate change" and everything to do with mismanaged forests and grasslands created by federal regulations on federal lands. This is where junk science and 'scientifically managed resources' bump heads."

The next day, Larson quoted this statement on both his Northwest and National programs and we made arrangements for me to be interviewed on the following day.

For many years I have been interviewed by Larson on his radio shows. These interviews usually take place during a wildfire season and focus on their cause, effects, and mitigation. We have also discussed other topics, such as spotted owl "habitat" and the management and ownership of the Elliott State Forest, but mostly our discussions have focused on wildfires—and usually as they were taking place, or shortly thereafter.

This year's Paradise Fire was the focus for both his regional and national shows on November 14, while the fire was still burning. The following transcript is edited from our conversation to highlight the historical magnitude of this fire, why it occurred, and what can be done to minimize future such occurrences.

Lars Larson: We've been talking a lot over the last week about the fires that began last Thursday, the fires in several places in California. But the most serious, one of the most deadly, in Paradise, California, just up the hill in the Sierra Nevada from Chico, California, is the one that's now identified as having killed almost fifty people. Several hundred people are still missing and unaccounted for. And the question is, what caused these fires? Or what could have made this situation better? The president sounded off by blaming the state forestry management, and I think there may be some blame there, but I think most of it is federal forestry management.

But I thought I'd consult with a guy who actually knows the subject well. Dr. Bob Zybach is a forest scientist, president of NW Maps Co. and the author of *The Great Fires: Indian Burning and Catastrophic Forest Fire Patterns of the Oregon Coast Range, 1491–1951*, his doctoral work that actually covered about five hundred years. Dr. Zybach, good to have you back on the program.

Bob Zybach: Thank you, Lars, good to be here.

Larson: Is there anything that would have made a difference in how the fires turned out for Chico, for Paradise, California?

Zybach: Oh, absolutely. If the fuels had been managed reasonably as had been done for thirty or forty years by the Forest Service, there would have been a lot fewer fires, including the Paradise Fire. And they would have been of a lot less magnitude, a lot safer, a lot less damage, and a lot more tenable.

Larson: Now when you say manage the fuels, you're talking about the wood mass that's grown up in those forests, and of course, the brush and the grass in the areas that are mostly federal lands around Paradise, California—the same kind of federal lands that we have in much of the northwest.

Zybach: Yeah, last year we had ten major fires in Western Oregon. All ten started on federal land. This year, we got the Klondike fire, which is the fourth major fire to come out of the Kalmiopsis Wilderness since 1987. Between 1952 and 1987, there was one fire of this magnitude in western Oregon—but one fire in thirty years when the land is being actively managed, compared to four fires coming out of the Wilderness, or ten fires all occurring on federal land last year are the patterns that are just the same this year.

Larson: Now you might say, I know some people are going to say, "Well, the government can only do so much." But let's go back about five hundred years and talk about how the Indians, who didn't have vehicles, who didn't have any of the rest of this, they managed these forests, and they did it sensibly and actually in a way that kind of mimics nature, did they not?

Zybach: Well, they managed the forests. I don't know about "mimicking nature." Some people think that's an area devoid of humans. But they used broadcast burning and fuel wood gathering, of course, to regularly burn tens of thousands of acres, sometimes to harvest crops, sometimes for hunting, probably sometimes for fun, or occasionally even by accident. But that kept the underbrush and the flash fuels under control. It kept fewer stems per acre so the canopies didn't fill in in most areas so densely. So it couldn't get crown fires in many areas. And then a lot of the land such as that in Sacramento Valley or the Willamette Valley was oak savanna. So it was easy to harvest, say, the tarweed in this area, or acorns, or bracken fern by having large controlled burns for thousands of years.

Larson: Now, Dr. Zybach, I'm not suggesting. I know some people do, but I'm not suggesting at all that we go out and have fires or deliberately set fires. Or that when a fire does occur, that we just let it burn and say, "That's mother nature's way." I know there's some tree huggers who like that.

But I'm suggesting something different that actually could be a money and job and tax revenue maker if we did it. And that is, I think what you've suggested, and that is you go and you actively do some logging. You take some of the revenue from that, and you do some thinning, and you do some clearing, and maybe you do some controlled burns at the right time of the year when there's enough moisture that you don't get this kind of massive fire. Is that what you're talking about? If not, please correct me.

Zybach: No, that's exactly right. In the sixties and seventies, we didn't have these fires, and we brought in hundreds of millions of dollars off of federal lands through logging. Maybe they were poorly designed logging units by today's standards, but it kept fires down. It brought hundreds of millions of dollars into rural communities and to rural families, and it produced lots of building materials for the housing boom, the baby boomers, and so on. So there were a lot of good, parks, schools in rural communities that came from the income. So this was tax-producing operations rather than firefighting which is an entirely different industry and dependent almost entirely on taxes to support it.

Larson: So, Doctor, tell me this. Is there any scientific reason not to go back to that same kind of logging activity, while maybe altering a bit the way it's done so that everybody can be happy and feel warm and fuzzy about it?

Zybach: There's zero scientific reason for not doing so. There are, however, a lot of so-called scientific rationales, whether it's global warming or black-backed woodpecker habitat or marble murrelet habitat. There's excuses and rationales and, I would say, junk science to support these approaches. But biologically, there's no problem at all with active management. And economically, it's wonderful. As far as danger to humans, we've just had a relatively few die or be injured, which is really tragic—but think of the millions of wildlife, animals that are killed in these events.

Larson: My understanding is the pictures of that coming out of the Paradise area of California are actually horrific. So Governor Jerry Brown, the lame duck governor of California, came right out almost immediately and said, "This is global warming. This is the new normal." Is there any scientific basis for that?

Zybach: No, it's a political basis. The climate's been the same for five hundred years. The weather events in which these fires occur are normal, east winds. It's just the fuel buildup and the management of those fuels, specifically on federal lands and specifically related to federal regulations of the last fifty years that have predictably resulted in these events.

Larson: So when Governor Brown says, "This is something new," as though Northern California's never had these north and east winds, as though Northern California has never been a very dry place much of the year except in the springtime when there's some rain. But most of Northern California is as dry as Eastern Oregon, maybe dryer, even to the point where there's less rain than there is evaporation. So it's very dry. But is that anything new based on the numbers?

Zybach: Nope, totally expected, normal, predictable, average. However you want to look at it, what's new is the fuel buildups: first from the elimination of Indian burning, second from the elimination of massive grazing. In the 1930s, we had what were called the "fern burners." That's one reason we had so many fires in the thirties, because it was a normal practice of people grazing sheep and other livestock in Oregon to fire the prairies after the grazing had been done to rejuvenate them for the following year. So that creates fire breaks and reduces opportunities for a catastrophic fire unless you have a prolonged drought and you have other fuels building up in adjacent lands, which is what also happened in the thirties.

Larson: The other thing I was intrigued by was your description of what a town is when a town is built the way most towns in the West are built, and that is, we build buildings out of stick lumber. We build them out of dimensional lumber that's dried in kilns. So it arrives relatively dried to the building site and then gets even drier after you put the building together as what little moisture is in it goes away. How did you describe it again?

Zybach: I think a "dismembered forest of bone dry firewood" was one of the ways I've described it. But it's a forest that's been essentially moved onto prairie lands. Almost all of the major towns in Western Oregon and Northern California are built on Indian prairies. They weren't forest; it wasn't cleared—it was predominantly prairie land. Then you move all these structures into places.

One of the things they're blaming the fires on is people moving into the rural area adjacent to forests and grasslands. It's government, so they give it an acronym, the "WUI" (pronounced "whoo-ee"). Then they create the "wild lands." So those are new designations for flammable areas, poorly managed Forest Service areas, quite often. They're saying all these people are moving into this acronym is the problem. Well, people have always lived there. But now it's 300,000 people or three million people instead of three hundred. They're bringing in massive amounts of firewood basically, structures built of flammable materials. That's what created the problem in Paradise and in Redding earlier.

Larson: The fires in Paradise actually started quite a distance outside of Paradise. As I understand, almost ten miles out of town. But it moved very, very quickly once the winds got hold of it.

Is there the potential, perhaps, in situations where you know you've got a town and you got a forest that's very dry that could catch fire, to put in firebreaks in the form of clear-cuts that would actually help protect a town by at least providing a defensible barrier between a town and the rest of the forest?

Zybach: Well, I think clear-cuts, that's nature's way. Whether it's a volcanic eruption or a wildfire or a Columbus Day storm, it's how nature regularly clears off hundreds of thousands of acres at a time, usually within a matter of less than a week, rather than clear-cutting, which might take years. But the real problem is the ladder fuels and the flash fuels in the savanna areas, such as around Paradise. That's an annual problem, or at least biannual.

Larson: The slash fuels are all that dried up buckbrush and grass and everything else that by this time of the year are bone-dry.

Zybach: Yep. That's why the fire moved so quickly. Then it gets into town with all that fuel. It's driven by an east wind, and the east wind is coming down off the Sierra Nevada, so it's warmed as it's getting lower in elevation and rapidly drying out the fuels in advance as well as creating its own weather patterns, which include winds that drive the fire further.

Larson: Dr. Zybach, anything you haven't said already that you would say if, say, Donald Trump put you in charge of the Forest Service. How you'd change things?

Zybach: We'd start making a lot of money, a lot of people would go to work, and there'd be a lot fewer catastrophic fires, a lot less smoke, and a lot more wildlife.

Larson: Unbelievable. Dr. Zybach, it's always a pleasure to have you on the program.

Zybach: Thank you, Lars.

(Used with permission)

(For more information by Dr. Zybach, check out this Web site http:// nwmapsco.com/ZybachB/Articles/Magazines/Oregon_Fish_&_ Wildlife_Journal/index.html.)

APPENDIX 7

Climate Change Reconsidered: Lists

List of Documents

This list of documents is a library of information on climate change by various authors. Each one is a study in itself.

Reports by the Nongovernmental International Panel on Climate Change (NIPCC) Thirteen reports spanning eleven years, 4,673 pages http://climatechangereconsidered.org/				
Title	**Lead Authors**	**Year**	**Pages**	**Link to digital PDF**
Nature, Not Human Activity, Rules the Climate	S. Fred Singer	2008	40	https://www.heartland. org/ template-assets/documents/ publications/22835.pdf
Climate Change Reconsidered	Craig Idso, S. Fred Singer	2009	856	https://www.heartland. org/ template-assets/ documents/CCR/CCR-2009/FullReport.pdf
NIPCC vs. IPCC	S. Fred Singer	2011	24	http://www.cfact.org/pdf/ NIPCCreport2011.pdf

Climate Change Reconsidered: 2011 Interim Report	Robert Carter, Craig Idso, S. Fred Singer	2011	416	https://www.heartland. org/ template-assets/ documents/CCR/ CCR-Interim/Full%20 Interim%20Report.pdf
Scientific Critique of IPCC's 2013 Summary for Policymakers	Robert Carter, Craig Idso, S. Fred Singer	2013	18	https://www.heartland. org/ template-assets/ documents/publications/ critique_of_ipcc_spm.pdf
Climate Change Reconsidered II: Physical Science	Robert Carter, Craig Idso, S. Fred Singer	2013	993	https://www.heartland. org/ template-assets/ documents/CCR/CCR-II/ CCR-II-Full.pdf
Chinese Translation of Climate Change Reconsidered	Robert Carter, Craig Idso, S. Fred Singer	2013	329	https://www.heartland. org/ template-assets/ documents/publications/ climate_change_ reconsidered-cn.pdf
Written Evidence Submitted to the Commons Select Committee of the United Kingdom Parliament	Robert Carter, Craig Idso, S. Fred Singer	2013	7	https://www.heartland. org/ template-assets/ documents/policy-documents/NIPCC%20 Testimony%20to%20 Parliament%20 re%20IPCC.pdf
Climate Change Reconsidered II: Biological Impacts	Robert Carter, Craig Idso, Sherwood Idso, S. Fred Singer	2014	1062	http:// climatechangereconsidered. org/wp-content/ uploads/2019/01/ CCR-II-Biological-Impacts-full-report.pdf

Commentary and Analysis on the Whitehead & Associates 2014 NSW Sea-Level Report	Robert Carter, W de Lange, J.M. Hansen, O. Humlum, Craig Idso, D. Kear, David Legates, Nils-Axel Mörner, C. Ollier, S. Fred Singer, Willie Soon	2014	44	https://www.heartland.org/publications-resources/publications/commentary-and-analysis-on-the-whitehead--associates-2014-nsw-sea-level-report
Why Scientists Disagree About Global Warming	Robert Carter, Craig Idso, Sherwood Idso, S. Fred Singer	2015	110	https://www.heartland.org/ template-assets/documents/Books/Why%20Scientists%20Disagree%20Second%20Edition%20with%20covers.pdf
Global Warming Surprises: Temperature data in dispute can reverse conclusions about human influence on climate	S. Fred Singer	2017	6	https://www.heartland.org/ template-assets/documents/policy-documents/Singer%20Global%20Warming%20Surprises.pdf
Climate Change Reconsidered II: Fossil Fuels	Roger Bezdek, Craig Idso, David Legates, S. Fred Singer	2019	768	http://climatechangereconsidered.org/climate-change-reconsidered-ii-fossil-fuels/

Books for Further Reading

The book you have just read just hit the high points. Read some of these following great books to get more:

Human Caused Global Warming: The Biggest Deception in History by Tim Ball

Through Green-Colored Glasses: Environmentalism Reconsidered by Wilfred Beckerman

Climate of Corruption by Larry Bell

Scared Witless: Prophets and Profits of Climate Doom by Larry Bell

Apocalypse Not: Science, Economics, and Environmentalism by Ben Bolch

Environmentalism Gone Mad: How a Sierra Club Activist and Senior EPA Analyst Discovered a Radical Green Energy Fantasy by Carlin, Alan

Climate, Man and History by Robert Claiborne

Eaten: A Novel by Susan Crockford

State of the Polar Bear Report 2017 by Susan Crockford

Population Bombed! Exploding the Link Between Overpopulation and Climate Change by Pierre Desrochers, et al.

Sustainable: The WAR on Free Enterprise, Private Property and Individuals by Tom DeWeese

Eco-Imperialism: Green Power, Black Death by Paul Driessen

50 Simple Things Kids Can Do to Save the Earth by Earthworks Group

Taken by Storm: The Troubled Science, Policy, and Politics of Global Warming by Christopher Essex, et. al.

Not by Fire but by Ice: Discover What Killed the Dinosaurs...and Why It Could Soon Kill Us by Robert W. Felix

Confessions of an Eco-Warrior by Dave Foreman

Energy and Environment, vol. 20, no. 1 and 2, by Bob Foster

Hot, Flat, and Crowded: Why We Need a Green Revolution—and How It Can Renew America by Thomas L. Friedman

An Inconvenient Truth by Al Gore

Earth in the Balance: Ecology and the Human Spirit by Al Gore

Our Choice: A Plan to Solve the Climate Crisis by Al Gore

The Assault on Reason by Al Gore

The Mad, Mad, Mad World of Climatism: Mankind and Climate Change Mania by Steve Goreham

Principles of Air Quality Management by Roger Griffin

Conservative Victory: Defeating Obama's Radical Agenda by Sean Hannity

Storms of My Grandchildren: The Truth About the Coming Climate Catastrophe and Our Last Chance to Save Humanity by James Hansen

A Primer on CO_2 and Climate by Howard C. Hayden

The Rough Guide to Climate Change: The Symptoms, the Science, the Solutions by Robert Henson

Power Grab: How Obama's Green Policies Will Steal Your Freedom and Bankrupt America by Christopher Horner

The Bottomless Well: The Twilight of Fuel, the Virtue of Waste, and Why We Will Never Run Out of Energy by Peter Huber, et al.

Climate Change Reconsidered II: Biological Impacts by Craig D. Idso

Climate Change Reconsidered II: Physical Science by Craig D. Idso

CO_2, Global Warming and Coral Reefs: Prospects for the Future by Craig D. Idso

Why Scientists Disagree About Global Warming by Craig D. Idso

Cool It: The Skeptical Environmentalist's Guide to Global Warming by Bjorn Lomborg

Global Crises, Global Solutions by Bjorn Lomborg

The Skeptical Environmentalist by Bjorn Lomborg

Scare Pollution: Why and How to Fix The EPA by Steven Milloy

Lukewarming: The New Climate Science that Changes Everything by Patrick J. Michaels, et al.

Meltdown: The Predictable Distortion of Global Warming by Scientists, Politicians, and the Media by Patrick J. Michaels

The Satanic Gases: Clearing the Air about Global Warming by Patrick J. Michaels, et al.

The Hockey Stick Illusion: Climategate and the Corruption of Science by A.W. Montford

Climate Change: The Facts by Alan Moran (ed.)

The Politically Incorrect Guide to Climate Change by Marc Morano

Break Through: From the Death of Environmentalism to the Politics of Possibility by Ted Nordhaus

Coming Climate Crisis? Consider the Past, Beware the Big Fix by Claire L. Parkinson

Heaven and Earth: Global Warming, the Missing Science by Ian Plimer

Environmental Overkill: Whatever Happened to Common Sense? by Dixy Lee Ray, et al.

Addicted to Energy by Elton B. Sherwin

Beyond Fossil Fools: The Roadmap to Energy Independence by 2040 by Joseph Shuster

Unstoppable Global Warming: Every 1,500 Years by Fred S. Singer

Going Green by Chris Skates

The War Against Nuclear Power by Eric N. Skousen

The 5000 Year Leap: A Miracle that Changed the World by W. Cleon Skousen

The Deniers: The World-Renowned Scientists Who Stood Up Against Global Warming Hysteria, Political Persecution, and Fraud, and Those Who Are Fearful to Do So by Lawrence Solomon

Climate Confusion: How Global Warming Hysteria Leads to Bad Science, Pandering Politicians and Misguided Policies That Hurt the Poor by Roy Spencer

The Great Global Warming Blunder: How Mother Nature Fooled the World's Top Climate Scientists by Roy Spencer

Landscapes & Cycles: An Environmentalist's Journey to Climate Skepticism by Jim Steele

Climate Change Denial: Heads in the Sand by Haydn Washington, et al.

But Is It True? A Citizen's Guide to Environmental Health and Safety Issues by Aaron Wildavsky

When All Plans Fail: Be Ready for Disasters by Paul R. Williams

ABOUT THE AUTHOR

 Jim Hollingsworth is a graduate of Pensacola Christian College with a master's degree in biblical studies and undergraduate degree in social science, with other works in geology and weather. He worked for nine years at the Bunker Hill Mine and has a strong background in climatology. He is a member of the CO_2 Coalition and the Global Warming Realists.